INTRODUCING
The Universe

Felix Pirani • Christine Roche

Edited by Richard Appignanesi

Icon Books UK ◆ Totem Books USA

This edition published in the UK in 2006 by Icon Books Ltd., The Old Dairy, Brook Road, Thriplow, Cambridge SG8 7RG email: info@iconbooks.co.uk www.introducingbooks.com

Sold in the UK, Europe, South Africa and Asia by Faber and Faber Ltd., 3 Queen Square, London WC1N 3AU or their agents

Distributed in the UK, Europe, South Africa and Asia by TBS Ltd., Frating Distribution Centre, Colchester Road, Frating Green, Colchester CO7 7DW

This edition published in Australia in 2006 by Allen & Unwin Pty. Ltd., PO Box 8500, 83 Alexander Street, Crows Nest, NSW 2065

Previously published in the UK and Australia in 1999

Originally published in the UK and Australia in 1993 under the title *The Universe for Beginners*

Reprinted 1996, 1997

This edition published in the United States in 2006 by Totem Books Inquiries to Icon Books Ltd., The Old Dairy, Brook Road, Thriplow, Cambridge SG8 7RG, UK

Distributed to the trade in the USA by National Book Network Inc., 4720 Boston Way, Lanham, Maryland 20706

Distributed in Canada by Penguin Books Canada, 90 Eglinton Avenue East, Suite 700, Toronto, Ontario M4P 2YE

ISBN-10: 1-84046-761-4
ISBN-13: 978-1840467-61-1

Text copyright © 1993, 2006 Felix Pirani
Illustrations copyright © 1993 Christine Roche

The author and artist have asserted their moral rights.

Originating editor: Richard Appignanesi

No part of this book may be reproduced in any form, or by any means, without prior permission in writing from the publisher.

Printed and bound in Singapore
by Tien Wah Press

THE UNIVERSE FOR BEGINNERS

The Universe is everything that exists.

Cosmology is the scientific study of the Universe.

More than can be seen. Beyond the Earth's atmosphere, only the Sun, the Moon, a few thousand stars, half a dozen planets, a few nebulae*, sometimes a comet, can be seen with the naked eye. Astronomers augment the naked senses with telescopes, spectrographs, radio dishes, tanks of cleaning fluid. They place instruments on rockets and satellites, and deep underground. Their 'seeing' is not the same as everyday seeing.

*There is a Glossary of unfamiliar and technical words, beginning on page 170

Ground rules for this book: The Universe is as it is because it was as it was, not because Anybody made it so.

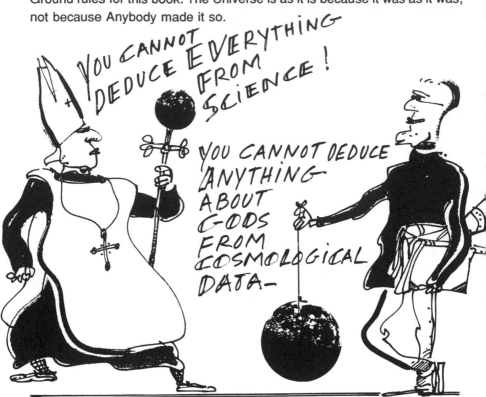

As far as this book is concerned, the Universe is a physical system, not a creation of the gods. It was not designed by a Great Designer. The sciences for studying it are physical sciences: astronomy and physics. It makes sense to ask

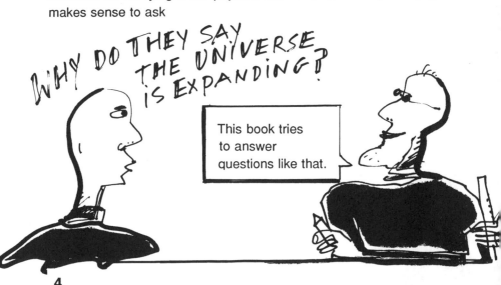

This book tries to answer questions like that.

WHY WAS THE UNIVERSE CREATED?

This is not a question for physical science. Questions of this kind demand intentional or teleological explanations that specify the future state which it is intended to bring about.

WHY ARE YOU BUILDING A TELESCOPE?

BECAUSE I INTEND TO PHOTOGRAPH GALAXIES

A star does not have intent. A galaxy does not act in order to bring about any result.

The Universe has a history. Stars form, evolve, radiate energy and change size. Galaxies spread apart: the Universe expands. You do not see stars or galaxies as they are now, but as they were when they emitted the light which is seen now.

outward in space =

Looking Backward in Time

Light from the Andromeda galaxy, for example, takes about 2,300,000 years to reach the Earth, so that Andromeda is seen, here and now, as it was about 2,300,000 years ago. The distance light travels in a year is called a light-year (ly); Andromeda is about 2,300,000 light-years away.

All things are made of atoms, little particles that move around in perpetual motion. The stars are made of atoms of the same kind as those on the Earth.

Which may seem obvious enough. Four hundred years ago few people believed it.

The atoms are clumped into things: molecules, living cells, organisms, rocks, mountains, oceans, planets, stars, galaxies. The Universe has structure. Here a planet, there a star; between, a little gas and dust.

Stars have structure of their own. And stars are clumped into galaxies, galaxies into clusters. The atoms themselves have structure. From atomic nuclei to clusters of galaxies there is a range in size of forty powers of ten or so.

It is the structure which makes the Universe interesting. Who would care about a Universe consisting of nothing but thin soup? Nobody, for in such a Universe there would be nobody to care about anything.

Putting the numbers in.

The exponent notation is very useful for writing powers of ten: If n is any whole number then 10^n means 1 followed by n zeroes and 10^{-n} means 1 divided by 10^n. For example 10^9 is a billion (1,000,000,000, a thousand million) and 10^{-6} is one millionth (1/1,000,000).

Powers of ten may be multiplied together by adding exponents: for example

$$10^9 \times 10^{-6} = 10^3.$$

Some particular powers of ten have special names:

Giga (G)	*means*	10^9
Mega (M)	*means*	10^6
Kilo (k)	*means*	10^3
Milli (m)	*means*	10^{-3}
Micro (µ)	*means*	10^{-6}
Nano (n)	*means*	10^{-9}

So that, for example, 3 Gly is 3×10^9 light-years, and 5 µm is 5×10^{-6} metres.

Later, 10^{-43} seconds comes in. 10^{-43} seconds is a tenth of a millionth of a millionth of a millionth of a millionth of a millionth of a millionth of a second.

The scale of the Universe

Imagine the scale of the Universe reduced by a factor 10^{10}.

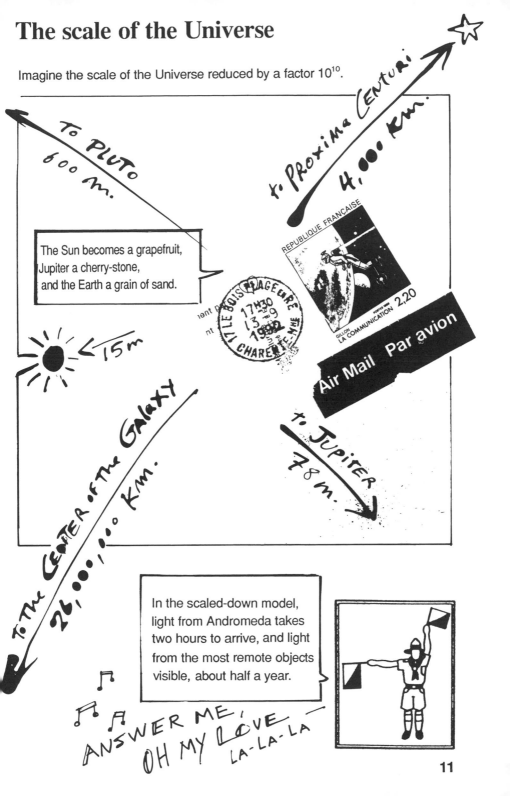

To PLUTO 600 m.

to PROXIMA CENTURI 4.000 Km.

The Sun becomes a grapefruit,
Jupiter a cherry-stone,
and the Earth a grain of sand.

← 15m

TO THE CENTER OF THE GALAXY 26.000.000 Km.

to JUPITER 78 m.

Air Mail Par avion

In the scaled-down model,
light from Andromeda takes
two hours to arrive, and light
from the most remote objects
visible, about half a year.

ANSWER ME, OH MY LOVE — LA-LA-LA

Cosmology is different from other sciences. There is only one Universe, but there are lots of cells, mountains, and stars.

— you cannot examine the Universe from outside, as you can examine a cell or a mountain.

little chance of experiment from here

I ASK You TOM—
HOW DO YOU KNOW
WHAT STARS ARE MADE OF—
HOW DO YOU KNOW
HOW BIG THE GALAXIES ARE—
WHAT DO YOU MEAN, THE UNIVERSE EXPANDS?

Virginia Woolf

THE ANSWERS
CHANGE, MY DEAR,
COSMOLOGY TOO
HAS A HISTORY

T.S.Eliot

Do I dare
Disturb the Universe?
In a minute there is time
For decisions and revisions which a minute will reverse.

Egyptian Cosmology

The sun-god Ra, having created himself, was united with his own shadow and begat twins, Shu, god of the air, and Tefnut, goddess of rain. Shu and Tefnut, united, also produced twins, the earth-god Geb and the sky-goddess Nut. Geb and Nut in turn were united, but this angered their jealous grandfather, Ra. Ra ordered Shu to separate them, and to hold Nut high above the Earth, as befitted a sky-goddess. Since then, Nut touches the Earth only with the tips of her fingers and toes. Her belly, covered with stars, which are her children, forms the arch of heaven.

The Greek connection

Western cosmology begins with the philosophers of Miletus, an ancient Greek colony in Asia Minor. Thales, traditionally the earliest of them, was reputedly the first to abandon mythical versions of cosmology and work out a secular picture of the Universe.

THE EARTH is A FLAT Disc FLOATING ON WATER - IN SOME WAY, EVERYTHING COMES FROM WATER -

THE SUN, MOON AND STARS ARE INCANDESCENT VAPOUR -

Thales may have been a practical man, too.

URSA MINOR! WE'RE LOST!

Anaxagoras of Clazomenae was first in the West to state clearly that

The MOON SHINES BY THE LIGHT iT RECEiVES FROM tHE

THE SUN iS A WHiTE-HOT STONE - NOT A DiVINITY--.

For this he was prosecuted for impiety. The Athenian leader Pericles made a speech in his defence.

GiVE THE LAD A BREAK- HE MEANS WEll...

BUT, FOR MY SiNS, I, ANAXAGORAS, WAS EXiLED TO LAMPSACUS...

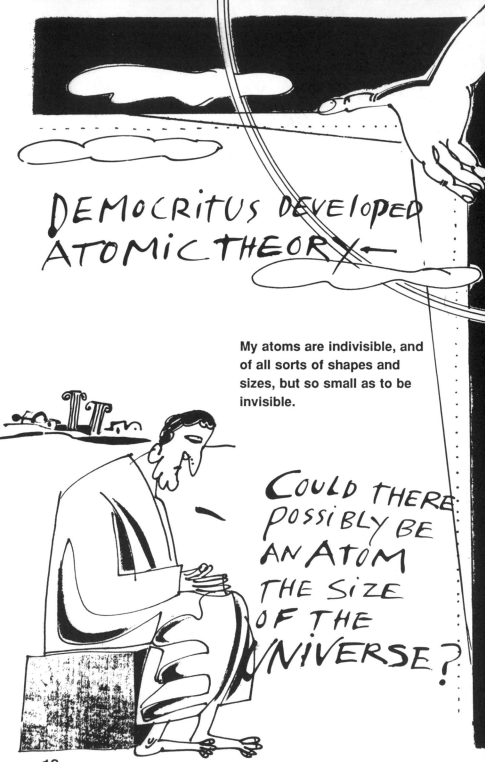

DEMOCRITUS DEVELOPED ATOMIC THEORY ←

My atoms are indivisible, and of all sorts of shapes and sizes, but so small as to be invisible.

COULD THERE POSSIBLY BE AN ATOM THE SIZE OF THE UNIVERSE?

For Plato the stars were living beings, divine, eternal. He maintained that the actual motions of the heavenly bodies must be resolved into uniform circular motions. This dictum prevailed in astronomy until the end of the sixteenth century.

Plato: GOD is ALWAYS BUSY WITH GEOMETRY—

Aristotle taught that the Universe has always existed. According to Aristotle, the Earth is spherical, and at rest at the centre of the Universe. The Universe is large but finite, and bounded by the sphere of the fixed stars, the primum mobile, with nothing outside. Within the primum mobile are nested spherical shells, all centred on the centre of the Earth, and all consisting of aether, the heavenly element, which is a crystalline solid.

Aether is an invisible, weightless substance, purer than fire. The planets and stars themselves are made from it. The planets, the Sun and the Moon are all embedded in aetherial shells. The shell containing the Moon is the lower boundary of the celestial region.

The motion of the inner shells is driven through mechanical linkages by further shells from the primum mobile, which is the source of all motion in the Universe. Altogether 55 spheres are required for the complete mechanism.

For Aristotle, each element has its natural position. The natural motion of the Earth as a whole, like that of each of its parts, is towards the centre of the Universe; that is the reason why it is now lying at the centre.

This notion of a fixed central Earth is inseparable from Aristotle's theory of mechanics. A moving Earth would also break down the absolute distinction which he drew between celestial and terrestrial matter. Worse still, if the Earth is merely a planet, the prospect of a plurality of worlds, with other inhabited planets, is opened up.

For the practical navigator, the stars hardly move relative to one another in the sky. But the stellar sphere moves daily around the Earth, and the Sun, the Moon and the five planets known in ancient times, Mercury, Venus, Mars, Jupiter and Saturn, move relative to the stars.

Vitruvius: Imagine seven ants crawling around circular channels on a potter's wheel. The wheel turns in the opposite direction. The ant which is nearest to the centre will finish its journey soonest, while the ant which goes round the outer edge of the wheel, with a longer circuit, will be much slower in completing the course, even if it move just as quickly as the others. In the same way the planets struggle on against the course of the firmament.

To describe the planetary motions, first Apollonius of Perga and then Hipparchus developed a new mathematical scheme, which was brought to its highest state by Ptolemy.

In the simplest form of the scheme, each planet moves steadily round the circumference of a circle, the epicycle, whose centre in turn moves steadily round the circumference of another larger circle, the deferent, whose centre is fixed at the centre of the fixed Earth.

Later the scheme was elaborated, with epicycles on epicycles on deferents. It lasted until the time of Copernicus — about 1400 years.

The Biblical Universe

The Christian doctrines, and even the Jewish doctrines which preceded them, have always been statements about spiritual reality, not specimens of primitive physical science. *C.S.Lewis*

The Reverend Harry Emerson Fosdick: In the Scriptures the flat earth is founded on an underlying sea; it is stationary; the heavens are like an upturned bowl or canopy above it; the circumference of this vault rests on pillars; the sun, moon and stars move within this firmament. There is a sea above the sky, and through the "windows of heaven" the rain comes down. This is the world view of the Bible.

The Chinese scholar Ko Hung was one of those who recognized the possibility of an infinite, largely empty Universe.

The Sun, the Moon and the company of stars float freely in empty space, moving or standing still, and all of them are nothing but condensed vapour.

Looking up at the heavens is like seeing yellow mountains sideways at a great distance, for then they all appear blue; or when we gaze down into a valley a thousand fathoms deep, it appears sombre and black. But the blue of the mountains is not a true colour, nor is the dark colour of the valley really its own.

The Jains, adherents of a South Asian religion, have numbers of gods, in various heavens, but they do not believe that any of these gods created the Universe. They think that the Universe has always existed and could not be destroyed.

Some foolish men declare that a Creator made the World.
This is not a good idea. Have nothing to do with it.

If a god made the World, where was he before he made it?
If he needed nothing to stand on, where is he now?

If no one was needed to make the stuff that the World was made from, then the World could have come about by itself, without the need for anyone to make it.

If he created the World out of love for living things and need for them, why did he not make the World completely happy and free of misfortune?

The idea that the World was created by a god makes no sense at all.

27

Christianity in the Middle Ages

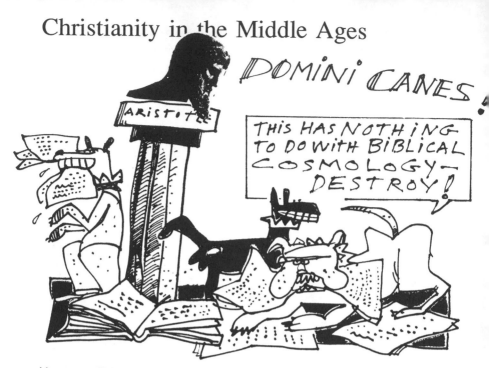

He was still being condemned at the beginning of the thirteenth century, but within a few years the Dominican friar Thomas Aquinas and others had embarked on a reconciliation.

Genesis says that God created the Universe. How can Aristotle say that it always existed?

The Bible uses metaphors which are addressed to the ignorant.

Aristotle's cosmology became dominant. Dante Alighieri celebrated it in his epic poem *The Divine Comedy*.

* WATCH DOGS OF GOD: A PUN ON DOMINICANS.

Copernicus in the year of his death

De Revolutionibus Orbium Cœlestium was published with a low-profile preface by the Lutheran theologian Andreas Osiander.

Moreover, Copernicus adhered to the Ptolemaic system of epicycles and deferents, so that his innovation does not yield as much simplification as it might have done.

Copernicus's important contribution was to displace the Earth from the centre. The idea that the Earth is not in a specially favoured position is sometimes called the Copernican principle. It leads naturally to the idea that the Earth is in a typical position in the Universe, which is the cosmological principle.

Asserting that the Earth is in a typical position amounts to saying that the Universe looks much the same from anywhere — to asserting its homogeneity.

This assumption is essential to modern cosmology

In 1572 the astronomer Tycho Brahe discovered a new star, or ···

NOVA! AS BRIGHT AS VENUS!

Novae are stars which become visible, or more noticeable, to observers on Earth as the result of a sudden increase in brightness. This one was probably what is now called a Type I supernova.

This nova must lie beyond the planets. And some comets are further away than the Moon. These discoveries put in question the unalterability of the heavens. I shall abandon Aristotle's crystalline spheres. But I don't need the motion of the Earth for my version of the Solar System, whatever Copernicus says.

THEY SAY HE LOST HIS NOSE IN A DUEL ···. SOME STAR!

The theory of planetary motion was enormously simplified by Johannes Kepler, who was assistant to Tycho and his successor as imperial mathematician to Rudolf II, the Holy Roman Emperor. Kepler worked on a mass of planetary data accumulated by Tycho. At first he followed the tradition that all planetary motions must be built of circles, but after laborious calculations he concluded that

Mars and other planets move in ellipses at variable rates.

This rejection of Plato's dictum was a major step towards modern astronomy.

Tycho, like most astronomers before him, seems to have believed in astrology. Kepler also made some money as an astrologer. He cast at least 800 horoscopes.

Kepler's laws

1. Each planet moves in an ellipse, with the Sun at one focus.

2. The radius vector from the Sun to the planet sweeps out equal areas in equal times.

3. The ratio

$$\frac{cube\ of\ the\ semi\text{-}major\ axis\ of\ the\ ellipse}{square\ of\ orbital\ period\ of\ the\ planet}$$

is the same for all the planets.

WHAT'S AN Ellipse?

An oval. You can draw one like this, with two drawing pins and some string. Each of the points where the pins are is called a *focus* (plural: foci).

WHAT's THE RADIUS VECTOR?

The line from the focus (where the Sun is) to the planet. It traces out the shaded area. When the planet is nearer to the Sun, it has to go further to trace out the same area in a certain time, so it goes faster. When it's further way, it goes slower.

SEMI-MAJOR AXIS

WHAT'S THE ORBITAL PERIOD?

The time it takes the planet to go all the way around once. The further away from the Sun a planet is, the longer it takes to go round. The Earth takes a year. Jupiter, which is about 5.2 times as far from the Sun, takes about 11.86 years (and 5.2^3 is about equal to 11.86^2, in agreement with Kepler's third law).

I can show that the MILKY WAY consists of STARS.

Galileo Galilei, working in Padua, first turned a telescope on the stars in 1609. He quickly discovered the satellites of Jupiter, the mountains of the Moon, sunspots, and the phases of Venus. The last of these discoveries was overwhelming evidence that Venus's orbit was centred on the Sun, but did not by itself prove anything about the Earth's orbit.

Domini Canes strike again...

We forbid the teaching of Galileo's discovery of sunspots at the University of Louvain.
And Innsbruck.
And Pisa.
And Salamanca.

The Aristotelian distinction between terrestrial and celestial had been obliterated. The perfection of the Sun had been put in question.

Father Clavius

Perhaps I can reconcile the new telescopic observations with the Aristotelian theory of the perfect sphericity of the moon. Maybe the valleys and mountains of the moon are covered by a crystalline substance, absolutely transparent, which is distributed so as to give the Moon a perfectly smooth surface.

Galileo was determined to persuade the Church authorities that they must come to terms with the new Copernican astronomy. But some Dominicans were out to get him. In 1616 the experts of the Inquisition were required to consider two propositions:

1 that the Sun is the centre of the Universe and hence is immovable

2 that the Earth is not the centre of the Universe, nor motionless, but moves, and turns daily on its axis.

The Experts:

The first proposition is foolish and philosophically absurd, and formally heretical inasmuch as it expressly contradicts many passages in Holy Scripture.

The second proposition receives the same censure in philosophy and is declared at least erroneous in faith.

The censors suspended the books of Copernicus, until "corrected". Books that taught the same doctrine were prohibited. Galileo was at that time protected by the powerful and Catholic Medici family, and his works were not specifically listed in the condemnation.

Cardinal Bellarmino :

I warn him that the doctrine attributed to Copernicus, that the Earth moves round the Sun and that the Sun is fixed in the centre of the universe without moving from west to east, must not be defended or held.

Galileo persisted in his attempts to drag the Catholic Church into the modern age.

In 1632 he published, with the Church's imprimatur, his *Dialogue concerning the two chief world systems*. At the time, the Pope, Urban VIII Barberini, previously a Galileo supporter, was in the midst of coping with a complex political crisis.

39

...and ...re... ...ord Cara... ...ito...
...mmonwealth, having ...fore my ...
...always believed, do now believe, and ...
...hed by the Holy Catholic and Apostol...
is ...wfully given to me by this Holy Offic... ...lly
...e of the World and immovable, and ...
...st not hold, defend, or teach the said f...
...een notified that the said teaching is c...
...the ... ...demned d...
...tting ...solut... ...hese...
...av...read and ...ved th...
...t the centre and m... Therefore, ...
...ristian this vehem... suspicion justly conce...
...curse, and dete... th... aforesaid error...
...contrary to ... Church; a...
...or in writt... ...that ...
...heretic or ...erson susp...
...of ...
...

...my said prom...
...ments... ...gated by ...
...h delinquen... ...
...ilea Galilei, have abjured, sworn, prom...
...the present document of my abjuration ...nd
Sopra Minerv... ...Rome, th... ...ne,

Galileo wins, 359 years late

AFTER 359 years, the Catholic Church has admitted that it was wrong to condemn Galileo (left) for asserting that the earth orbits the sun.

At the weekend the Pope accepted the results of a commission of the pontifical academy of sciences he set up 13 years ago to study the case.

The Inquisition condemned Galileo in 1633 for backing a theory of the astronomer Copernicus because it clashed with the Bible.

"Better late than never," said Margherita Hack, director of astronomy of the University of Trieste. — Reuter.

The development of astronomical instruments and astronomical theory proceeded hand-in-hand during the eighteenth century. With the Industrial Revolution came improved methods of glass-making and metal-working, both important for astronomy. Towards the end of the century, William Herschel constructed larger and more powerful telescopes than had been possible before.

The planet had been recorded a score of times before by various astronomers, but none had noticed its distinctive appearance.

I called the planet *Georgium Sidus* after George III, King of England. The King made me his private astronomer, at a salary of £200 a year, and later added another £50 for my sister Caroline, who helped me with my work.

NEVER MIND THE HALF DOZEN COMETS I'VE DISCOVERED.... Silly ME!

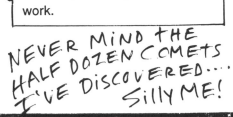

In the early days of the 40-foot telescope, we all got up from dinner and went into it, and sang "God save the King", with accompaniment on the oboe and any other instruments we could get hold of.

Just the same, everybody calls the planet Uranus, and not Georgium Sidus.

The development of astronomy continued in the nineteenth century, but it was not clearly understood by astronomers until early in the twentieth century that

the Sun is a star of rather common type, located within the Milky Way.

Nor was it known until the 1920s where the boundaries of the Milky Way might be drawn.

The spiral nebulae had not been definitely located beyond the confines of the Milky Way, nor had individual stars been distinguished in them.

Shortly after World War I, a reflecting telescope with a 2½ metre mirror, at that time much the largest and the most powerful in gathering the light of faint objects, was installed on Mount Wilson in California. With its help Harlow Shapley sketched a map of the Milky Way Galaxy, and located the Sun in its outer parts.

The 'spiral nebulae' are also galaxies. They are systems of stars similar to the Milky Way but much more remote. *Edwin Hubble*

Without reason to prefer one galaxy above another, one could no longer appeal to astronomy to prescribe a centre for the Universe.

The Milky Way

The Milky Way, the Galaxy in which the Sun is a star, is visible to the naked eye as a long bright cloud in the sky. Galileo discovered in 1610 that it is made up of a great number of stars. There are several times 10^{11} stars in the Galaxy, which is shaped rather like a fried egg sitting in a cloud of smoke. Its mass is about 10^{12} times the mass of the Sun.

The bulge in the middle of the Galaxy — the yolk of the egg — is about 32 thousand light-years from one side to the other. The surrounding white of the egg is called the *disk*; it is about 130 thousand light-years across but only about two thousand light-years thick. The disk is probably made up of spiral arms, but this is not absolutely certain. The Sun lies in the disk, about 26 thousand light-years out from the centre of the bulge.

The cloud-of-smoke shape is called the *halo*. Not much of it can be seen, but it may be five or six hundred thousand light-years from one side to the other. There are stars in the halo, but it is not known what the rest of it is made of.

The stars in the Galaxy appear to be rotating around its centre; it takes the Sun about 250 million years to go once around.

What's between the stars?
Gas and dust. Mostly hydrogen. In this part of the Galaxy between 1 and 10^8 hydrogen atoms per litre.

How many atoms in a litre of water?
About 10^{26}.

thin stuff
that stuff between
The stars —

There may be from about a million to about 10^{13} stars in a single galaxy.

A galaxy may be
elliptical: rugby-ball-shaped
spiral: Catherine-wheel-shaped
irregular: shaped like nothing in
particular.

Most galaxies are found not by themselves but in groups or *clusters* of tens, hundreds or thousands. The Milky Way belongs to a group called the Local Group, with at least 25 members.

The Local Group is about 3 million light years across.

The nearest big cluster, called the Virgo cluster (because it is seen in the same direction in the sky as the star constellation Virgo) is probably between 50 and 70 million light-years away. Distances to remote objects are very hard to determine with accuracy.

The Virgo cluster is thought to be about 7 million light-years in diameter, and has over 1200 known member galaxies.

The galaxies in each cluster appear to move around the cluster. Most clusters are grouped together. Both the Local Group and the Virgo cluster belong to the Local Supercluster, which is pancake-shaped, about 100 million light-years in diameter and 10 million light-years thick. It has about 100 member clusters.

Many galaxies appear to lie on the surfaces of bubble- or sponge-like structures. The bubbles may be 100 million light-years or so in diameter.

There seems to be no end to the grouping of galaxies together. We have found a group of thousands of galaxies spread out in an immense crumpled sheet. We call it the Great Wall It is at least 500 million light-years wide. *Margaret Geller*

The extent of the largest features is limited only by the size of the survey, which so far covers only one hundred-thousandth part of the visible Universe. *John Huchra*

The expanding Universe

Hubble's formula for the speed of recession *v* of another galaxy at distance *r*

$$V = H_0 r$$

which means that the speed increases with distance.

Double the distance

and you double the speed of recession of that galaxy, as seen from this one, or of this one, as seen from that.

The number H_0 is called Hubble's constant. It is not precisely known, because it is very difficult to determine the distances between galaxies with precision. H_0 is currently thought to be close to 22 kilometres per second per million light-years.

What happens if you run the expansion backwards?

The galaxies must have been closer to one another in the past than they are now. At one time, their contents must have been squashed close together and the appearance of the Universe must have been very different.

If the laws of nature which are believed to apply now have always applied to the Universe, then there were initially no stars or galaxies but only a rapidly expanding hot gas of elementary particles — hot because it was compressed — out of which stars, galaxies and other structures subsequently developed. The further back you go, the hotter and denser do the contents of the Universe become. At least that is the majority view.

One idea was that the Universe started its life a finite time ago in a single huge explosion and that the present expansion is a relic of the violence of this explosion.

But the name *BIG BANG* caught on.

The name, Big Bang, is unfortunate because it may be misunderstood as referring to an event at a *singular* start of expansion of the Universe.
James Peebles

BANG

Present-day physics has nothing to say about how the expansion started.

There are mathematical models — Friedmann models, explained later — in which the Universe expands from an initial *singularity*. There is no reason to suppose that this correctly represents the real Universe.

In any case, current physical theory cannot deal with a "beginning". The earliest that the theory can deal with is a time at which the contents of the Universe were compressed to the so-called Planck density, around 10^{93} times the density of water. When "big bang" is used here, it means the time at which the Universe was in that state. The Planck density is explained later.

*There is no evidence for a **beginning** of the Universe, and no theory can deal with it adequately.*

The big bang idea ran into a difficulty.

Observations made in the 1930s and the 1940s led to the conclusion that the present ages of the Earth, the Sun and the Galaxy were greater than the time since the big bang. The time since the big bang seemed to be about 2 billion years. But there was evidence that the Earth was about 4½ billion years old. How could the Earth have been in existence since before the big bang?

BUT ONE DAY AT THE MOVIES —

A group known as the Cambridge Circus proposed the steady state solution.

In the steady state model of the Universe, the Universe expands, and new matter is created in the gaps left by the expansion at just such a rate as to maintain a roughly uniform distribution of matter, "steady" — of unchanging average density — forever.

The details change continually, but the overall appearance of the Universe is always the same.

In this model there is no big bang; running the Universe backwards or forwards one always sees the same average overall appearance.

The Cambridge Circus

We argued that continual creation of matter in small quantities in an always existing Universe...

...was no more of an assumption than big bang creation of Universe and matter all at once.

Moreover, it brings the process into the realm of the testable.

BUT THE STEADY STATE THEORY FAILED The TEST.

58

The difficulty about the time scale persisted until 1952, when Walter Baade, working at Mount Palomar Observatory, revised the scale of distances to other galaxies.

Baade: the time since the big bang is likely to have been greater than five billion years — long enough for the evolution of Galaxy, Sun and Earth to have taken place.

Interest in the steady state model declined, and almost vanished after the discovery in 1965 of the cosmic background radiation ...

...EXPLAINED LATER

How old is the Universe?

What do you mean, "How old is the Universe?" I just told you that it may not have had a beginning. It could have been going on forever.

You could ask: how long since the big bang? That depends on how fast the Universe is expanding. If the Universe is expanding fast, then there has been less time since the big bang. If the Universe is expanding slowly, more time.

They can't tell exactly how long it is since the big bang. But much more precise estimates are possible now than used to be the case. The time since the big bang is thought to be close to 13.7 billion years.

In 1965, Arno Penzias and Robert Wilson, two radio astronomers working at Bell Laboratories in New Jersey, were tracing sources of radio noise.

They had to clean the pigeon droppings out of the antenna first.

George Gamow had predicted that about

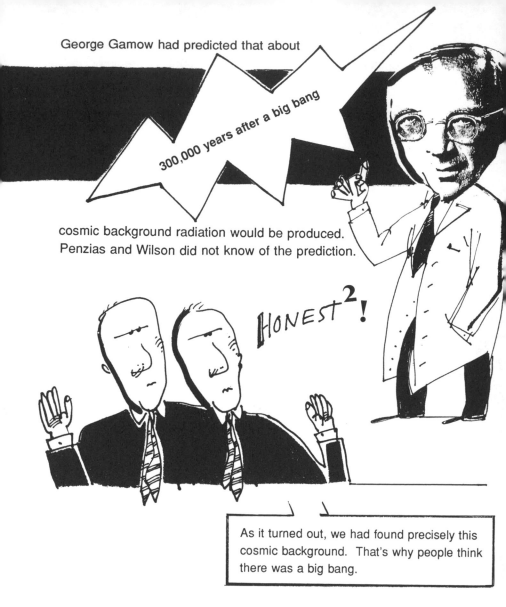

300,000 years after a big bang

cosmic background radiation would be produced.
Penzias and Wilson did not know of the prediction.

HONEST^2!

As it turned out, we had found precisely this cosmic background. That's why people think there was a big bang.

Since the work of Penzias and Wilson, almost all cosmologists now accept the idea of the big bang, meaning an early, highly condensed state from which the expansion of the Universe may be supposed to have proceeded. Remember that "big bang" does not mean the beginning of the Universe. Physics has nothing to say about a "beginning".

Laws of nature

Modern cosmology begins with some basic assumptions.

☆

The rest of the Universe is made of atoms of the same kind as those on the Earth.

☆

The same laws of nature apply in the rest of the Universe as on the Earth.

☆

The same laws apply to the past as to the present.

✪

The idea of a 'law of nature' goes back at least to the seventeenth century French philosopher René Descartes.

I suppose that God rules the Universe entirely by 'laws of nature' which were decided upon at the beginning.

In contrast, during the middle ages God participated in the day-to-day running of the Universe.

Perhaps the idea of specifically quantitative 'laws of nature' came from the international associations of merchants, like the old Hansa League, who had laws of their own, which dealt with numbers, weights and measures.

Eugene Wigner: **It is not at all natural that "laws of nature" exist, much less that man is able to discover them.**

In modern science 'laws of nature' are ways of formulating the regularities of natural phenomena.

The Laws of Modern Physics

The relevant laws are laws of physics, embodied in two fundamental theories which describe the behaviour of matter and radiation: general relativity and quantum mechanics.

Einstein General relativity is primarily a theory of gravitation on the large scale — the scale of stars and planets, and upwards.

Schrödinger and Heisenberg Quantum mechanics is primarily a theory of phenomena on the small scale — the scale of atoms and their nuclei.

There are four fundamental forces: gravity, electromagnetism, and the 'strong' and 'weak' nuclear forces.

Energy is conserved: it may be converted from one form to another, but there is no process known in which it is created or destroyed. However energy can manifest itself as mass, with the result that sometimes conversion of energy from one form to another may appear as the conversion of mass into energy or vice versa, according to Einstein's formula $E = mc^2$. This is the case when stars radiate energy, as you will see later.

Under the moderate conditions which prevail on the Earth, most matter exists in the form of atoms. Each atom has a central nucleus (the plural is nuclei), surrounded by a cloud of electrons. Nearly all the mass of the atom resides in the nucleus. Atoms combine to form molecules; in a pure chemical compound every molecule is made up of the same combination of atoms.

For example every molecule of water consists of two atoms of hydrogen and one of oxygen.

"Hydrogen" and "oxygen" are the names of chemical elements. To understand what a chemical element is, you need to know something about the structure of atoms.

The forces which hold the electrons in an atom are electromagnetic forces. Responsiveness to electromagnetic forces is measured by electric charge, which can be positive or negative. The nucleus of an atom is charged, and so are the surrounding electrons. By convention, the charge on the nucleus is called positive, while the charge on an electron is called negative.

Opposite charges attract, *like charge repel —*

To begin with, one may think of an atomic nucleus as made up of protons and neutrons (protons and neutrons are thought to be made up of still more basic particles called quarks). A conventional unit of electric charge is the charge on the proton: a proton has charge +1. An electron has charge -1, and a neutron has no charge. An atom by itself — a neutral atom — has zero total charge. The number of electrons in the cloud around the nucleus is the same as the number of protons in the nucleus itself, which is called the atomic number.

ELEMENTS

Hydrogen	1
Azote	5
Carbon	54
Oxygen	7
Phosphorus	9
Sulphur	13
Magnesia	20
Lime	24
Soda	28
Potash	42

Atoms which have the same atomic number — the same number of protons in their nuclei — are atoms of the same element. The chemical behaviour of an atom is determined almost completely by the atomic number. Atoms with atomic number 1 are atoms of hydrogen, those with atomic number 2 are helium, ... those with atomic number 92 are uranium.

Nuclei of a particular element need not all have the same number of neutrons. For example helium nuclei may have one neutron or two, as well as the two protons which they all have. An isotope of an element is a particular kind of nucleus of that element with a particular number of neutrons.

Protons and neutrons are called nucleons. The number of nucleons in a nucleus of a particular isotope is called the nucleon number. Ordinary hydrogen has one proton and no neutrons; its nucleon number is 1. There is a hydrogen isotope called deuterium, which has one proton and one neutron, so that its nucleon number is 2. The nucleon number is often written as a superscript to the conventional abbreviation of the name of an element: hydrogen is H, ordinary hydrogen is ^{1}H, and deuterium is ^{2}H. Hydrogen also has an isotope ^{3}H, called tritium. Helium is abbreviated He, and the common isotope of helium is ^{4}He.

The forces which act between nucleons are called nuclear forces. Two different nuclear forces are recognised, called "strong" and "weak".

The strong force holds the nucleons together in the nucleus, overcoming the electrostatic repulsion between the protons. The weak force is involved in the decay of neutrons: a free neutron, not ensconced in any nucleus, decays into a proton, an electron, and a third particle, like the neutron not electrically charged, called a neutrino.

Energy for stars and H-bombs

At sufficiently high temperatures, which occur naturally in the interiors of stars and are produced artificially in hydrogen bombs, light nuclei fuse together to form heavier ones. In a series of steps, four ^{1}H nuclei fuse to form a ^{4}He nucleus. The mass of a ^{4}He nucleus is slightly less than four times the mass of a ^{1}H nucleus.

slightly less, master?

The "slightly less" is released as energy according to Einstein's formula $E = mc^2$.

Such processes account for generation of the energy radiated by stars, for the production of other chemical elements, and for the force of the H-bomb.

Nearly all of what humans know of the rest of the Universe they know from reception of electromagnetic radiation.

Visible light,
Radio waves,
Infra-red and ultraviolet radiation,
X-rays and gamma rays.

X RAY TUBE

In the 1860s James Clerk Maxwell completed the unification of the theories of electricity and of magnetism, which had been in progress for about half a century.

The theory of electromagnetism describes the relations between electric and magnetic fields, and their production by electric charges and currents. The existence and properties of electromagnetic radiation, which is produced by moving charges, may be deduced from these equations:

$$\nabla \times \mathbf{E} + \frac{\partial \mathbf{B}}{\partial t} = 0 \qquad \nabla \cdot \mathbf{D} = \rho$$

$$\nabla \times \mathbf{H} - \frac{\partial \mathbf{D}}{\partial t} = \mathbf{J} \qquad \nabla \cdot \mathbf{B} = 0$$

The different kinds of electromagnetic radiation may be distinguished by frequency. The radiation consists of oscillating electric and magnetic fields; the frequency is the rate of oscillation. It is measured in cycles per second, or Hertz (Hz), named after the physicist Heinrich Hertz.

All electromagnetic radiation travels in empty space with the speed of light. In the Earth's atmosphere, the speed is slightly lower, in a denser medium, still lower.

I have produced radio waves and verified that they travel with the same speed as light.

HEINRICH HERTZ

A Megahertz (MHz) is a million Hertz. The radio waves from an FM station at 100 on the dial have a frequency of 100 MHz. To every frequency there corresponds a wavelength. 100 MHz radio waves have a wavelength of just under 3 metres.

Waves or particles?

Depending on the circumstances, radiation may be perceived as waves, sharing some physical characteristics of sound waves and water waves...

Radio waves

$$10^8 \quad 10^7 \quad 10^6 \quad 10^5 \quad 10^4 \quad 10^3 \quad 10^2 \quad 10 \quad 1 \quad 10^{-1} \quad 10^{-2}$$

or as particles, sharing some of the characteristics of atoms, molecules and bullets.

Particles of electromagnetic radiation are called photons.

Frequency and wavelength are related by the formula:
frequency × wavelength = c.

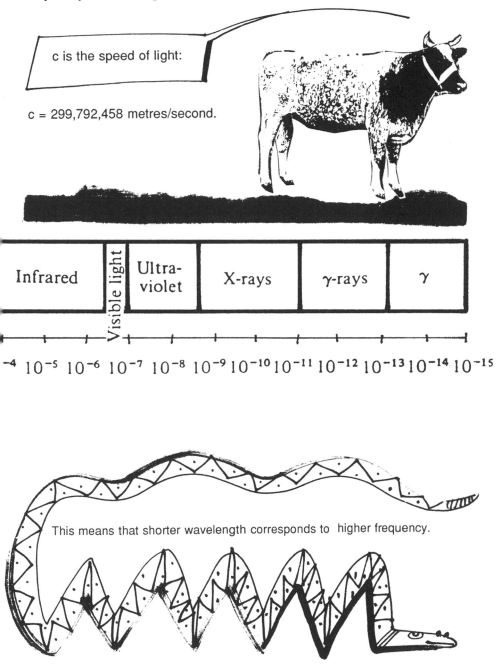

c is the speed of light:

c = 299,792,458 metres/second.

Infrared	Visible light	Ultra-violet	X-rays	γ-rays	γ

$^{-4}$ 10^{-5} 10^{-6} 10^{-7} 10^{-8} 10^{-9} 10^{-10} 10^{-11} 10^{-12} 10^{-13} 10^{-14} 10^{-15}

This means that shorter wavelength corresponds to higher frequency.

Visible light

The wavelengths of visible light lie between about 0.4 and about 0.66 μm (μ is explained on p 10). A range of wavelengths or frequencies of light or other radiation is called a spectrum (plural: spectra). Visible light is spread out into a spectrum by water droplets in the atmosphere, which produces the rainbow. The colours always appear in the order Red, Orange, Yellow, Green, Blue, Indigo, Violet, which is the order of decreasing wavelength, or increasing frequency of the radiation.

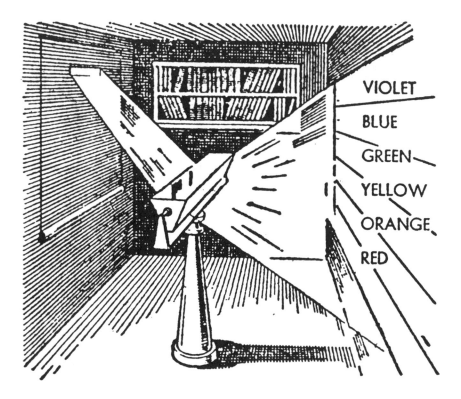

Light may also be spread out artificially into a spectrum, for example by passing it through a prism. The intensity of the radiation will vary over the spectrum. How the intensity varies depends on how the radiation is produced, so if you know the intensity variation of a source of radiation, this gives you evidence about what the source is like.

Spectra

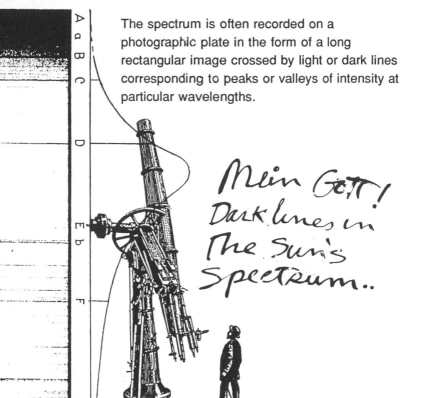

The spectrum is often recorded on a photographic plate in the form of a long rectangular image crossed by light or dark lines corresponding to peaks or valleys of intensity at particular wavelengths.

Mein Gott! Dark lines in The Sun's Spectrum..

Joseph von Fraunhofer

Kirchhoff's Rules

Each chemical element has its own characteristic line spectrum. Three rules: (1) a dense hot gas produces a continuous spectrum, containing a whole range of wavelengths; (2) a tenuous hot gas gives rise to a bright line spectrum (emission lines); (3) in front of a source of light with a continuous spectrum, a cool gas absorbs some light and produces a dark line spectrum (absorption lines).

Redshift — Blueshift

What is known of other parts of the Universe is known from the study of electromagnetic radiation they emit. That's how Hubble discovered the expansion of the Universe. Absorption lines in the spectrum of light emitted by a receding object are observed to be systematically shifted towards longer wavelengths; this is called a redshift, because the longest wavelengths of visible light are red, and what is visible appears shifted towards the red (wavelengths beyond the visible are also shifted).

Similarly the light emitted by an approaching object is systematically shifted towards shorter wavelengths; this is called a blueshift, because the shortest wavelengths of visible light are blue, and what is visible appears shifted towards the blue.

In this respect light waves resemble sound waves: if the source of the waves is receding from you, the waves will seem to you to be longer than if the source is at rest. If the source of the waves is approaching you, the waves will seem to you to be shorter. In the case of sound waves it is this effect that produces the drop in pitch of an ambulance siren as the ambulance passes you.

... ON THE OTHER HAND — IF YOU HAPPEN TO BE LYING IN HERE ...

Austria 1842 — Dear Diary — It goes without saying that in the case of light waves, the whole spectrum is shifted

Christian Doppler.

Special relativity

Towards the end of the nineteenth century it became apparent that Maxwell's electromagnetic theory could not be fitted into the same framework as Newton's mechanics.

A new framework was needed. Something had to give.

This new framework was special relativity theory, published by Albert Einstein in 1905. It combines space and time into a unitary "space-time", and gives special significance to the speed of light.

Space-time diagrams

1.

What are dimensions? Space has three dimensions: length, breadth and thickness.

At the Earth's surface North-South, East-West, and up-down.

2.

A photograph is a two-dimensional representation of a three-dimensional scene. A map has two dimensions (the piece of paper on which the map is printed has three dimensions, but is very limited in one of them compared to the other two).

Time and Space

3.

Time also may be counted as a dimension. A space-time diagram shows positions in time as well as in space. In this diagram, there is one space dimension, which runs left and right. Here and in the diagrams which follow, time runs upward.

4.

In most of these diagrams, only one space dimension will be shown. Each person, or particle, or planet has a world-line in space-time. In the diagram, your world-line is just a graph showing your movements. If you stay in one place, it goes straight up. Time passes, but there is no change of space position.

5. If you move to left or right, your world-line is inclined, to show that your position in space is changing.

6. The faster you move, the more it is inclined, because your space position changes more in the same time.

World-lines for light

If you speed up, your world-line
inclines more; if you slow down it
inclines less. So if your speed
varies, your world-line is curved.

A photon is a particle of light, so a light ray is
just the world-line of a photon. According to
special relativity, in empty space, nothing can
move faster than light. Radio waves and other
electromagnetic waves move at the same
speed as light.

Past and Future

9.

This space-time diagram, shows two space dimensions and the time, drawn in perspective. The light rays form a light cone, separated into two parts at the Here-and-Now point.

10.

Here are two world-lines, yours and A's. A is in the Andromeda galaxy. She sends you a radio signal.

For simplicity, you are both shown not moving. Actually, Andromeda does move relative to our galaxy.

11.

A is about 2,300,000 light-years away. You don't see her as she is now, but as she was, 2,300,000 years ago.

12.

If you send her a reply, it will take 2,300,000 years to reach her.

Looking outward in space = looking backward in time.

13.

The further away something is, the longer ago did it send out the light which you see.

14.

You can receive signals only from events within, or on, your past light cone.

AS TIME PASSES, MORE AND MORE OF THE UNIVERSE BECOMES VISIBLE TO YOU.

Newton's Law of Gravitation

The force of gravity keeps the planets moving in their paths around the Sun, and the satellites in their paths around the planets. This same force keeps the stars moving round in a galaxy, and the galaxies moving round in a cluster. These motions may be explained by Newton's law of universal gravitation.

Newton's law says that every object attracts every other object with a force proportional to their masses and inversely proportional to the square of the distance between them:

$$F = \frac{GMm}{r^2}$$

where F is the force, M and m the masses, and r the distance. The formula means that if you double either of the masses, it doubles the force; if you double the distance, it reduces the force to a quarter of what it was.

The law doesn't tell what the force is in any particular case. For that, you need to know the value of the constant G in front, which is called the constant of gravitation.

The value of G is approximately $G = 6.6726 \times 10^{-11}$ m^3/kg.sec^2

How Attractive are you?

The gravitational attraction between two 400,000-ton supertankers sitting side by side in drydock is about the same as the weight of an adult.

The gravitational attraction between two consenting adults...

Is roughly the same as the weight of a piece of typing paper 2mm square.

Mass and Weight

Mass is not the same as weight.

Mass is a measure of an object's resistance to being accelerated

—IT'S INERTIA.

Mass enters Newton's equation:

force = mass × acceleration.

If here you put in the value of the acceleration due to gravity, this equation becomes:

weight = mass × acceleration due to gravity.

The same object will have different weights in different places, although its mass remains the same.

The mass of a particular astronaut will be the same on the Earth, on the Moon and in space.

The weight will be smaller on the Moon, where gravity is weaker than on the Earth,

AND THIS IS ONE PLACE WHERE YOU DON'T CARRY MUCH WEIGHT, SWEETHEART—

Galileo showed by experiment that all bodies, or at least all he tested, fall equally fast, although he probably didn't do it by dropping weights from the top of the Leaning Tower of Pisa.

All bodies fall equally fast

Suppose that M is the mass of the Earth, and m the mass which is dropped from the Leaning Tower. Then the law of gravitation says that

$$F = \frac{GMm}{r^2}$$

and the law of motion says that

$$F = ma$$

where a means the acceleration. If the m which comes into the law of gravitation is the same as the m which comes into the law of motion, then for motion under gravity

$$ma = \frac{GMm}{r^2}$$

and the m may be cancelled out of this equation, leaving

$$a = \frac{GM}{r^2}$$

The acceleration is the same, whatever the mass.

My theory allows this
but doesn't require it.

In 1915, Einstein published his general relativity theory, a generalization of his special theory which was designed specifically to include gravitational forces. In his theory, all bodies must fall equally fast. They can't help it.

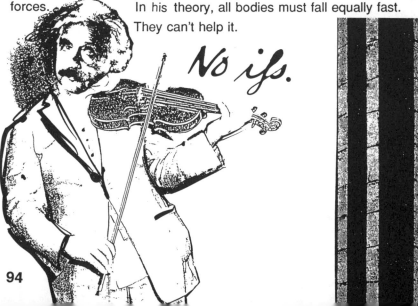

No ifs.

General relativity

My theory gives more accurate answers than yours for the motion of the planets.

The predictions are only very slightly different.

I get the motion of the planet Mercury right. You don't. For the Universe as a whole, my theory gives quite different answers.

General relativity is based on the principle of equivalence, which asserts that gravitational and inertial effects are indistinguishable.

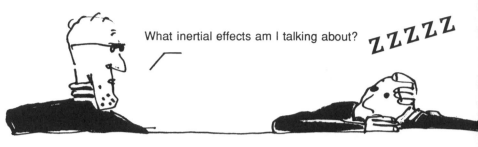

What inertial effects am I talking about? ZZZZZ

Acceleration, for example. Imagine yourself shut up in a space capsule. You can do experiments inside the capsule, but you can't see out. You feel that you're being pulled towards the floor of the capsule...

Is the capsule sitting on the Earth's surface, so that you feel the Earth's gravity...

or is it somewhere out in space, being boosted upwards by a rocket, so that you feel the acceleration?

I couldn't tell—

Now suppose that there isn't any pull towards the floor.

Then the capsule must be floating out in space, with the rocket motors turned off.

Couldn't it be falling down an elevator shaft towards the centre of the Earth?

Air resistance would slow it down, and I'd still feel a pull.

Suppose the air had been evacuated from the elevator shaft. Then you wouldn't feel any pull.

Then I couldn't tell whether the capsule was falling down the elevator shaft, or floating out in space.

That's the principle of equivalence.

NOW ABOUT YOUR INERTIA...

To accord with the principle of equivalence, general relativity describes gravitational effects and inertial effects in the same way, in terms of the "geometry" of space-time.

Which Means?

An ordinary map is flat. The space-time of general relativity is something like a map with the relief embossed on it. The geometrical shapes of the relief describe the gravitational field.

General relativity also predicts the existence of gravitational radiation, produced by moving masses. This is analogous to the electromagnetic radiation produced by moving charges in Maxwell's theory. Like electro-magnetic waves, gravitational waves carry energy, and travel with the speed of light. The effects of gravitational waves have been observed, although they have not yet been detected directly.

Everything attracts everything else by gravitational forces. For example, matter attracts photons. To put it another way, gravitation distorts light-cones. The effect of this is that light rays passing near a massive object seem to bend. The bending of light rays by the Sun, predicted by Einstein, was first observed during an eclipse in 1919.

Black holes

Stars attract the photons which they themselves have emitted.

If the gravitational field is strong enough, radiation cannot escape from the neighbourhood of the star.

The star is there, but it cannot be seen. A star like this is a black hole. Light and matter can go in, but nothing can come out.

You can look at this in another way: If you throw a ball up into the air, gravity pulls it back.

How high it goes depends on how fast you throw it, and in which direction.

If you propel a satellite into space with a rocket, gravity pulls it back. Whether it comes back to Earth...

goes into orbit, or leaves the Earth's neighbourhood for good depends on how fast you propel it, and in which direction.

The speed with which you need to propel an object straight up, in order that it can escape completely from the gravitational field, is called the escape velocity.

At the surface of the Earth, the escape velocity is about 11 km per second. At the surface of the Sun, it is over 600 km per second.

Some objects are thought to have collapsed to such a small size, for their masses, that the escape velocity is greater than the speed of light. These are black holes. The gravitational field is so intense that nothing can escape from it. No particles. No radiation. Once in, you stay in. That's why you can't see a black hole.

However, radiation can be emitted from the region surrounding a black hole by a process which has the effect of reducing the mass of the black hole.

This process is called Hawking radiation, after the theoretical physicist Stephen Hawking, who inferred its existence from quantum theory. For black holes formed by the collapse of stars, this process is too slow to be of much importance except on a very long time-scale.

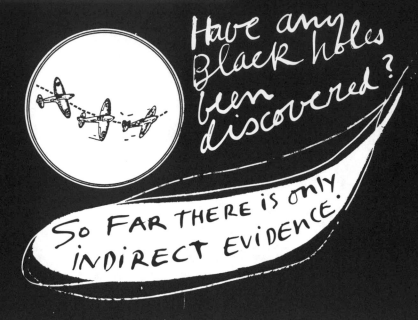

Have any Black holes been discovered?

So FAR THERE IS only INDIRECT EVIDENCE.

Stars with large enough masses may end up as black holes, and there may be very massive black holes at the centres of some galaxies.

So an ordinary star can become a black hole?

Astronomers think so. An ordinary star is a ball of gas held together by gravity, which radiates electromagnetic energy, generally energy produced by nuclear fusion of hydrogen into helium. Some energy is released also as the star contracts under gravity.

Not making it.

"Brown dwarfs" with less than about 8% of the mass of the Sun do not achieve nuclear fusion.

Small is BeautiFul.
No Chance For
Brown Dwarfs
Ever Becoming
Black Holes—

They can still emit electromagnetic radiation released by their contraction under gravity.

The most massive stars have masses about 150 times the mass of the Sun, and are several million times as luminous. The mass of the Sun is about 1.989×10^{30} kg.

The Sun won't last forever

The present phase of the Sun's existence is expected to last about 10 billion years altogether. In a star with 12 times the mass of the Sun, this phase lasts, according to theory, only a few million years.

The Sun, which is a medium-size star, is thought to have been formed about five billion years ago, and is expected to continue for about another five billion years in its present state. After that, it will become a red giant star, much bigger and rather cooler than it is now. It will swallow up Mercury and Venus, and everything on the Earth will be burnt up. If there are any people left on the Earth by then, they will have to find somewhere else to live. Some time after that, the Sun is expected to become smaller and hotter, a white dwarf star. A white dwarf with the mass of the Sun may be the size of the Earth.

Supernovae

The evolution of a star depends critically on its mass. The more massive the star, the more gravity compresses its material, the higher the central temperature becomes, and the more rapidly the nuclear reactions take place.

The subsequent history is also different. According to theory, a star with more than about 12 times the mass of the Sun eventually collapses under gravity, and then rebounds in an explosion, blowing off its outer layers and becoming very bright — a supernova. A supernova may become 10 billion times as bright as the Sun.

Neutron stars

If what remains after a supernova explosion has a mass more than about 1.4 times and less than about three times the mass of the Sun, then according to theory the remains become a neutron star: an object composed entirely of neutrons. A neutron star with the mass of the Sun would have a radius of only 20 km. At the surface of a neutron star, the escape velocity is around half the speed of light.

If the mass which remains is greater than about three times the mass of the Sun, then it seems inevitable, from what is known now, that the remains of the supernova become a black hole. But no black hole has been definitely identified yet.

HAS ANYBODY FOUND A NEUTRON STAR?

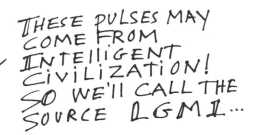

Jocelyn Bell, then a graduate student at Cambridge University, discovered a source of very regular radio pulses, one pulse every 1.3373 seconds.

THESE PULSES MAY COME FROM INTELLIGENT CIVILIZATION! SO WE'll CALL THE SOURCE LGM1...

...FOR LITTLE GREEN MEN NO 1, OF COURSE.

But I was soon to discover three more pulsating sources, and about 450 more have since been found. Now they are called pulsars, and they are generally thought to be rotating neutron stars which send out narrow radio beams. Without any little green men to steer them.

Cambridge

CLASS OF 67

Quasars

The other favourite black hole candidates are quasars.

In 1963 this puzzle was solved:

Maarten Schmidt: This spectrum is just the hydrogen spectrum, with an enormous redshift.

These can't be ordinary stars. Let's call them quasi-stellar radio sources.

Or quasars, for short.

Today, most astronomers believe that quasars are very distant Active Galactic Nuclei (AGNs, for short). The most central one percent or so of a galaxy may be an AGN.

The light of a galaxy isn't just the combined light of its stars, if it has an AGN. An AGN doesn't emit normal starlight. It may produce as much radiation as a hundred billion stars. As much as a whole ordinary galaxy. Powerful radiation over a great range of wavelengths.

Many astronomers think that the quasars in AGNs are supermassive black holes, maybe several billion times the mass of the Sun. Gas from the surroundings is attracted by the black hole, which compresses and heats it. This hot dense material emits radiation very intensely.

There is a black hole at the centre of our Galaxy, too.

It may be that some quasars are formed when two galaxies collide, and that small ones grow by sucking in gas from their surroundings. It seems that a few billion years ago, quasars were far more common than they are now. But none of this is guaranteed. Astronomers' ideas about black holes and quasars are very likely to change with time.

Quantum theory.

It all started with black bodies. A black body is different from a black hole. A black body is an object which absorbs all the radiation which falls on it, reflecting nothing.

There are no perfect black bodies, but for example the inside of an electric oven which has been running steadily for some time with the door closed is a good approximation to a black body. A small region deep inside the Sun, where the gas and radiation are almost in equilibrium, would be another good approximation.

Black bodies absorb all the radiation which falls on them, but they emit radiation, too. According to theory, when a black body emits radiation, the radiation has a characteristic variation in intensity over the spectrum, and this intensity is completely fixed once you know the temperature of the body.

Max Planck devised the original quantum theory about 1900 to explain the spectrum of a black body.

The energy of my model radiating body can be emitted or absorbed only in definite packets, which I call quanta. The energy in each quantum is proportional to its frequency: **Energy = h x frequency**

h is called Planck's constant:

$$h = 6.626176 \times 10^{-34} \text{ Joule seconds.}$$

In 1913, Niels Bohr, by assuming that atoms exist only in definite quantum states, extended Planck's idea to explain the spectrum of hydrogen.

112

In the 1920s, Erwin Schrödinger, Werner Heisenberg and others devised a more elaborate and powerful theory, called

QUaNTUM MECHANICS!

a theory generally applicable to atoms and their component parts and to radiation.

QUANTUM THEORY IS NOW THE FOUNDATION FOR THEORIES OF ELEMENTARY PARTICLES, ATOMS AND NUCIEI—

Putting it all together

Special relativity and quantum mechanics can be unified, but there is no satisfactory synthesis of general relativity and quantum mechanics. Physicists have been trying to meld them together for over 50 years, without much success.

Why is this synthesis needed?

Because neither general relativity nor quantum mechanics is by itself adequate to describe the Universe before the Planck time.

The Planck time

The Planck time is the earliest time at which present-day physics makes any sense, the time at which the density of the contents of the Universe was the Planck density.

The Planck density d_{Pl} and the Planck time t_{Pl} are defined in terms of three fundamental constants of nature, Planck's constant h, the gravitational constant G, and the velocity of light c: $d_{Pl} = c^5/hG^2$ and $t_{Pl} = \sqrt{(Gh/c^5)}$.

The Planck density is about 10^{93} times the density of water. Dense.

The Planck time is about 10^{-43} seconds. Short.

Present-day physics says nothing reliable about the nature of the Universe before the Planck time. Before the Planck time, the structure of space-time itself is uncertain.

Chris Isham: Our current understanding of space and time leaves much to be desired. There is no real evidence justifying the use of the Einstein equations [of general relativity] at subatomic scales.

This means that conditions for a beginning of the Universe cannot be specified theoretically. Cosmologists speculate about them anyhow.

Cosmologists make theoretical models — mathematical models — of the things they are trying to explain. Other scientists do the same.

The models are generally fairly simple. We incorporate what seem to be the most basic features of whatever is being modelled.

The most useful models are generally those which can be elaborated, to take account of more subtle features, without becoming too complicated.

Friedmann models.

I think I'll make some mathematical models of the Universe, starting from Einstein's theory. No details.

Alexander Friedmann

I did not know of these models before I discovered the Universe was expanding.

HONEST!

But Hubble's data could be fitted to a Friedmann model.

A FRIEDMAN MODEL IS LIKE THIN Soup —

Models mustn't be stretched too far. Every model is limited in validity. The Friedmann models expand, and give Hubble's formula, but tell you nothing about the formation of the galaxies.

HERMann Bondi

The Friedmann models do not present the detailed structure of the real Universe. A Friedmann model contains a completely uniform expanding (or contracting) distribution of structureless matter, without any stars or galaxies, let alone planets or people. The material of the Universe is there, but all the structure is totally smeared out.

119

Friedmann models and

If you were in a Friedmann model, the Universe would look simplest to you if you were to go with the flow — move together with the smeared out material.

Imagine hypothetical astronomers in a Friedmann model, who go with the flow, without it making any difference to the model. Let's call them co-movers, because they move with the flow. You can work out what they would see, immersed in the model with all their instruments, but not having any effect on the way it behaves.

Cosmic Time

The co-movers have a common time
standard. And in most, but not all,
Friedmann models there is an initial
singularity from which the model can be
said to begin.

Time measured from the initial singularity
by the common standard is called cosmic
time.

So in the mathematical model there may
be a beginning of the Universe. This
feature of the model cannot be used to
picture the real Universe. A Friedmann
model can have no validity before the
Planck time.

The cosmological principle.

Each Friedmann model is exactly homogeneous.

This means that what co-movers see does not depend on where they are. At any chosen cosmic time, the model looks exactly the same to all co-movers.

Each Friedmann model is exactly isotropic.

This means that what co-movers see does not depend on which way they look. At any chosen cosmic time, the model looks exactly the same in every direction, to any co-mover.

The cosmological principle asserts that the Universe is homogeneous, and isotropic around every co-mover.

According to the cosmological principle, every co-mover describes the history of the Universe in the same way, and makes no distinction between different directions in the model.

The Friedmann models satisfy the cosmological principle exactly. The real Universe, on the other hand, satisfies the cosmological principle only approximately, for it is full of structure. It is clumpy at least up to the scale of superclusters of galaxies. Remember the Great Wall?

On the largest possible scale, the distribution of matter and motion seems to be rather accurately isotropic, but cosmologists differ about the degree of homogeneity.

The largest scale is of the order of ten billion light-years, about as far as the most remote observable galaxies.

124

What Structure Do you Get?

Departures from isotropy and homogeneity develop during the expansion from the hot dense state, and stars and galaxies form.

Random motions of the galaxies, typically 500 km/sec, are superimposed on the flow.

The Universe remains on average homogeneous and isotropic. But the standard model is a lot more realistic than the Friedmann models.

The early Universe ⊕

Whatever started the expansion, perhaps an inflationary epoch, perhaps something more wild, is not intrinsic to the standard model.

According to the standard model, in the first seconds of the expansion the Universe is so hot that no atomic nuclei, let alone ordinary atoms or molecules, can subsist.

The standard model Universe cools as it expands. At a cosmic time of about one minute the synthesis of the lightest nuclei can begin. This process is called nucleosynthesis. Deuterium, helium and lithium are produced in the first thousand seconds.

As the expansion proceeds, the energy of radiation declines relatively and absolutely, and at a cosmic time of about 300,000 years, the formation of hydrogen atoms becomes possible.

matter   *radiation*

Subsequently the interaction between radiation and matter can to a great extent be neglected, and the remaining radiation travels on through the Universe practically without impediment. According to the standard model, it is this relic radiation which constitutes the cosmic background discovered by Penzias and Wilson. Because it was produced at a time when matter and radiation were practically in equilibrium, the relic radiation has a typical black body spectrum.

127

In the standard model, as the Universe continues to expand, galaxies and stars begin to form, largely under the influence of gravitational forces, and the Universe develops to its present state. The departures from homogeneity and isotropy become significant, but these properties are still, on a sufficiently large scale, preserved. The basic units of the Universe are galaxies.

There are plenty of problems, too. For example, there is no single theory of the formation of galaxies and larger structures which fits all the observations.

Cosmologists cannot even agree whether galaxies formed first, and then collected into clusters, or whether clusters formed first, and then broke up into galaxies.

The Future of the Universe

The standard model doesn't specify the amount of mass in a given volume — the density of matter. This has to be determined from observation. It makes a difference to what happens in the future.

If the contents of the Universe consist only of ordinary matter, then gravitational attraction slows down the expansion of the Universe. If the density is high enough, then gravitation will halt the expansion altogether and the Universe will, after some time, stop expanding and begin to collapse, reverting to a hot dense phase. Such a Universe is said to be **closed**.

This final phase would not be just a big bang in reverse, however, because the collapsing material would be clumped into stars, galaxies and clusters, and there would be a lot of black holes; in its early stages the expanding Universe is rather smooth and unclumped. The end point of a collapse is called a Big Crunch.

If the density is too low to halt the expansion, then the Universe will continue to expand forever. In this case the Universe would be said to be **open**.

In the borderline case, the Universe just continues to expand forever, but at a rate which becomes smaller and smaller. The density is said to be at its **critical value**.

The best available evidence suggests that the density is at, or very close to, the critical value. However, the rate of expansion is not becoming smaller and smaller. The expansion is accelerating, and has apparently been accelerating for the past 7 billion years or so. It is deduced from this that the Universe contains something else besides ordinary matter. What the something else may be is at present quite uncertain. Some possibilities are explained on page 141.

A big crunch

Suppose that the density of matter were so high that gravity could overcome the expansion. The Universe would continue to expand for some time from now, but at a decreasing rate.

After the expansion stopped...

As time went on the blue shift of each galaxy would increase. Ever more remote galaxies would appear blue-shifted.

As the Universe contracted, clusters of galaxies would merge into one another, and eventually galaxies themselves would merge. The present black body radiation, supplemented by the blue-shifted radiation from galaxies, would eventually become hotter and more intense than the radiation from stellar interiors. Everything would roast.

But that might not be the end. There might be a bounce: Some cosmologists suppose that the Universe oscillates. Instead of spreading out for ever from the big bang, or spreading out once from the big bang and then contracting once to a final big crunch, they conjecture that the Universe bounces back, each crunch turning into a new big bang. This is very speculative.

The far future of an open Universe

Freeman Dyson once worked out what would happen in the far future, if the Universe were open. If you assume that

> The laws of physics do not change with time.
> The relevant laws are already known to us.
> The Universe is open. Then...

1. Within 10^{14} years the longest-lived low mass stars will exhaust their hydrogen fuel and cool to very low temperatures. Stars of larger mass will reach their final states in shorter times.

2. After 10^{15} years, the Sun may be expected to have encountered another star closely enough to detach an Earth-like planet. The time until an encounter close enough for severe disruption of orbits will be considerably shorter.

3. After 10^{19} years, the central region of the Galaxy may be expected to collapse to a black hole, while stars in its outer regions are detached from it.

4. After 10^{20} years the Earth-Sun system would decay by radiation of gravitational waves. But the Earth will almost certainly escape from the Sun before gravitation pulls it in, unless the Sun escapes from the Galaxy with the Earth still in orbit.

5. After 10^{24} years stellar orbits around the Galaxy will decay by gravitational radiation.

6. After 10^{64} years a black hole of solar mass would decay by radiation. Decay of a black hole of galactic mass could take up to 10^{100} years. At the end of its life every black hole will become for a short time very bright. The cold expanding Universe will be illuminated by occasional fireworks for a very long time.

7. After 10^{1500} years all ordinary matter will have fissioned or fusioned to iron by radioactive processes.

What a way to go!

How they see it

"Yesterday all the past. The language of size
Spreading to China along the trade-routes...
Yesterday the shadow-reckoning in the sunny climates...
But today the struggle...
To-morrow, perhaps, the future:... **the gradual exploring of all the
Octaves of radiation;** W. H. AUDEN —

Richard Goodman (1939): the only poem by an Englishman anywhere
near being a revolutionary poem.

Auden (1964): This is trash I am ashamed to have written.

How do they compare the standard model, and all that goes with it, with the actual Universe?

The history of astronomy since 1945 has been dominated by the gradual exploring of all the octaves of radiation.

Radiation is absorbed by the Earth's atmosphere. The extent to which it is absorbed depends on the frequency. Visible light, and radio waves of frequencies between about 1.5 MHz and 300 GHz are impeded very little, but some other frequencies are almost completely absorbed before they reach ground level.

Until 1945 practically no astronomical observations were made except in visible light. Except for one Bell Telephone Laboratories researcher:

In 1933 Karl Jansky was searching for sources of interference with 20 MHz radio equipment being developed for ship-to-shore and transatlantic telephone communication.

HERE'S SOMETHING THAT SHOULD MAKE THE HEADLINES: I'VE DISCOVERED A SOURCE IN THE DIRECTION OF THE CENTER OF THE GALAXY!

The discovery made the front page of the New York Times.

BUT AT THE LABS:

ANY PROFIT HERE? NO? THEN STOP HIM!

BELL

During World War II:

We have discovered radiation emitted by the Sun at radio frequencies!

After the War there were both the scientists and the technology for a systematic exploration of the Universe at radio wavelengths.

And we can start observing from these captured German rockets.

In the 1970s it became possible to make observations from satellites. This removed all the limitations on astronomy which result from the absorption of incoming radiation by the Earth's atmosphere.

If society has a technical need, that helps science forward more than ten universities.

Friedrich Engels

The cosmic background.

The clearest evidence that the Universe is homogeneous and isotropic on a large scale is the cosmic background radiation, discovered by Penzias and Wilson.

The cosmic background radiation has a black body spectrum. This fits the theory that the radiation was produced in the early Universe when the contents were in a hot, dense condition and radiation and matter were in equilibrium.

Radiation with this kind of spectrum can be characterized by giving a single number, called its black body temperature.

The average black body temperature of the radiation is now about 2.7 degrees Kelvin (degrees above absolute zero). This temperature is found to be extraordinarily uniform. It does not vary by more than a few parts in 100,000, as the antenna detecting the radiation moves around, pointing to different parts of the sky.

Interference from other sources of radiation, and absorption by the Earth's atmosphere, impede accurate measurement of the background radiation from the ground. Hence the interest for cosmology in the Cosmic Background Explorer (COBE) satellite, the first satellite dedicated to measurements of the cosmic background radiation. It was designed, built and in 1989 launched by scientists working for the US National Aeronautics and Space Administration (NASA).

A spacecraft equipped with more accurate and sensitive high-resolution instruments, the Wilkinson Microwave Anisotropy Probe (WMAP), named in memory of a leading member of the COBE team, was launched in June 2001. Its microwave radiometers measured temperature variation of the Cosmic Microwave Background (CMB) with great precision.

The new data should also lead to great advances in understanding the development of galaxies.

Some Cosmologists maintain that...

This is the dawn of a new era ...

... the era of 'precision cosmology'.

The WMAP data lead to a scenario with a Universe 13.7 billion years old, with total energy density close to the so-called critical value (explained on page 129). From WMAP and other data it is concluded that only about 4.4% of the content of the Universe consists of ordinary matter. About 22% is **cold dark matter (CDM)**, and about 73% is **dark energy**. "Cold" just means that the CDM is slow-moving; recent observations suggest that it may be slightly warmer. The CDM clusters – it slowly collects under gravitational force to form clumps, which help the formation of stars, galaxies and other astronomical structures. The existence of the CDM is inferred from its gravitational effects. For example, stars in a spiral galaxy rotate around the centre of the galaxy. The speeds of rotation have been measured for hundreds of spiral galaxies. The centrifugal force ought to balance the gravitational force, but there is not enough visible matter to make this come out right. The CDM supplies this need.

The dark energy does not cluster. Its function is to maintain the large-scale geometry of the Universe and to mediate the speeding up of the expansion, which is thought to have been going on for about the past 7 billion years.

Neither the CDM nor the dark energy can be seen directly, and neither has been conclusively identified – which is to say that 95% of the matter in the Universe is of uncertain constitution. There are plenty of speculations about what it could be.

One conjecture has been that the dark energy is associated with a uniform but slowly-evolving low-energy physical field, sometimes called "quintessence". Another suggestion is that it is the "lambda-term" in the equations of general relativity, named after the Greek letter λ, which was originally introduced by Einstein for a completely contrary reason. This solution is quite unsatisfying, in any case, because it offers a source of gravitational field which is not related to any other physical phenomena – it just happens to be needed to fit the observations.

Einstein's theory may be wrong, or in need of major revision. Even if it is right, there are some other problems to solve.

the
H O R
P R ⊙

One cannot know anything of a past event unless some kind of signal could have been received from it.

Since no signal travels faster than the speed of light, there are events from which a signal cannot yet have arrived. You can know nothing about such events.

For example, you can know nothing about the Andromeda galaxy that happened less than about 2,300,000 years ago, because the Andromeda galaxy is about 2,300,000 light-years away.

There are distant events in well-separated parts of the sky which are so remote from one another that a signal cannot ever have been sent from one to the other.

space

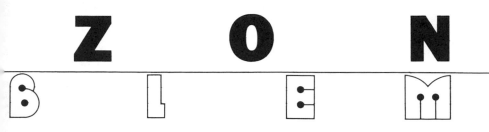

This raises a problem about the cosmic background radiation. Its black body temperature varies from point to point in the sky by only a few parts in 100,000.

But the cosmic background radiation is thought to have originated only about 300,000 years after the big bang. At that time points in the sky which are well separated, as you see them now from the Earth, cannot have been within signalling range of each other, so that no physical interaction can ever have taken place between them.

There is no possible physical mechanism by which they could have arrived at equal temperatures. The initial uniformity of temperature has to be put into the standard model as a separate assumption. This is the horizon problem.

143

Fine-tuning the density?

Another problem is presented by the density of matter.

The average density of matter in the Universe is not fixed by theory. The critical value, just low enough for the Universe to be able to go on expanding for ever, is an unstable one.

If the density ever differs slightly from critical, the difference increases with the passage of time.

From some knowledge of the present density, you can work the standard model backwards. It turns out that if the present value of the density is in the range considered probable, then at a cosmic time of one second it must have differed from critical by less than one part in 10^{15}, and earlier it must have been very much closer still to the critical value.

This "fine tuning" of the average density is thought by many cosmologists to be inherently implausible.

These problems are solved, at least after the Planck time, by **inflationary cosmology**, which has been developed from a form proposed by Alan Guth in 1981. In the inflationary scenario, an exceptionally rapid expansion of the Universe, which may have lasted no longer than 10^{-35} seconds, takes place after the big bang. This is the inflation.

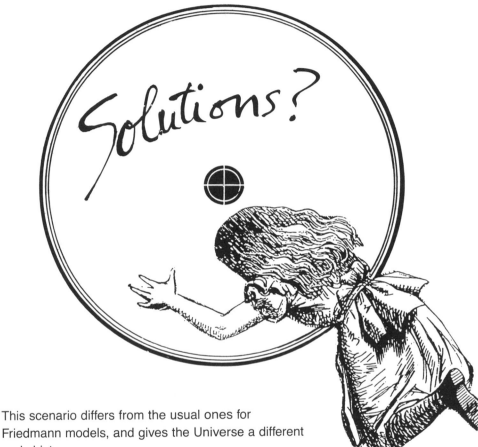

This scenario differs from the usual ones for Friedmann models, and gives the Universe a different early history.

Contrary to the usual case in the Friedmann models, the region which evolves into the presently observable Universe is such that all parts of it could have been in mutual communication for long enough to establish homogeneity and isotropy of the expanding material. This disposes of the horizon problem.

Inflation.

Many variants on Guth's original theory have been proposed subsequently. It soon became apparent that there were model Universes in which inflation would go on forever. Inflation would stop in some regions of space, forming what have been called mini-universes or "pocket universes". One of these would comprise all that can be observed by humans here and now.

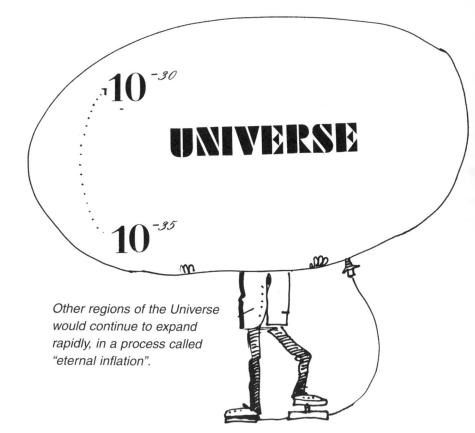

10^{-30}

UNIVERSE

10^{-35}

Other regions of the Universe would continue to expand rapidly, in a process called "eternal inflation".

One result of this would be that the pocket universes would not be able to communicate with one another. Under certain conditions, inflation would produce an infinity of these pocket universes, inhabitants of which would never be able to observe anything about each other. They would never be able to confirm or deny the existence of the other pocket universes.

In inflationary cosmology...

The cosmological principle is deduced, not assumed —

UNIVERSE

The isotropic property of the cosmic background radiation follows from the theory. The rapid expansion also forces the average density to be exactly critical.

The elementary particle theorists...

Here's an opening for us! We can work at far higher energies than are possible on Earth. The high temperatures in the early Universe are associated with such high particle energies.

WE'Re in Business!

There are more things in heaven and earth, Horatio, than are dreamt of in your philosophy.

A big problem which remains is the description of the Universe before the Planck time, as you were warned on page 114.

Physicists are convinced that the dense hot matter of the earliest Universe can be described reliably only with the help of both general relativity, which describes the gravitational field, and quantum mechanics, which deals with the behaviour of particles and fields on a microscopic scale. A quantum theory of gravity is what is needed.

Such a theory has been sought for over fifty years. Among current attempts to formulate a quantum theory of gravity, a favourite is **String theory**.

String theory

One of the essential capabilities of a quantum theory of gravity is that it be able to describe the properties of gravitational radiation, which plays for the gravitational field a role analogous to that played by electromagnetic radiation for the electromagnetic field. The particle of electromagnetic radiation is called the photon. The particle of gravitational radiation is called the **graviton**. A striking novelty of string theory is that it does not merely allow the introduction of the graviton – it forces it upon you.

When it was realised that string theory might be the long-sought-for Theory of Everything – incorporating gravitation as well as electromagnetism and nuclear forces and capable of describing all the fundamental particles – interest in it blossomed, although it amounted to a radical departure from conventional particle physics.

The basic idea of string theory is that fundamental particles of ordinary matter, including electrons, photons, gravitons, neutrinos and quarks, are not point-like as they are presumed to be in conventional physics, but one-dimensional extended objects, something like infinitely thin rubber bands, or rubber bands that have been cut once, which are called **strings**.

On average, strings are a Planck length (about 10^{-33} cm) long – far too small to be observed directly. The observed properties of the fundamental particles correspond to modes of vibration of the strings.

One of the major difficulties encountered in the construction of a quantum theory of gravity arises from Heisenberg's Uncertainty Principle, which is fundamental to quantum theory. It follows from the Uncertainty Principle that the smaller the scale on which a physical measurement is attempted, the stronger the quantum fluctuations that will impede this attempt. A consequence is that investigation of space-time geometry at distances below the Planck scale is effectively impossible; in a theory with point particles, Einstein's theory breaks down at Planck distances when quantum theory is taken into account. However, string theory effectively smears out the fluctuations at the smallest scales without invalidating gravitation theory.

This coherence of gravitation and quantum theory is achieved at a price: in relativity theories, space-time is 3+1-dimensional, with three space dimensions and one time dimension (see page 83). In the starter version of string theory, the Universe must have six extra space dimensions, so that space-time is 9+1-dimensional. In the sophisticated version, explained later, there are seven extra, and space-time is 10+1-dimensional. The extra dimensions are curled-up, and so small that they could not be apparent to measuring instruments.

This is not the first time that a physical theory of nearly everything with extra dimensions has been proposed. In the 1920s, Theodor Kaluza and Oskar Klein developed a 4+1-dimensional theory unifying gravitation and electromagnetism. Some would say that string theory is a relative, perhaps a descendant, of Kaluza–Klein theory.

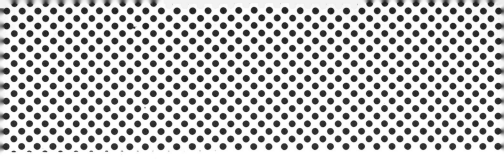

In quantum theory, it is very often the case that a physical prediction is a prediction of probabilities of certain possible results. In order for this to make sense, the predicted probabilities must be positive, and with the conventional scaling of probability they must lie between 0 and 1. In the working out of string theory, it was found that unless space-time was endowed with the six extra dimensions, some probabilities turned out to be **negative**. So the extra dimensions are essential. But the space-time we live in is 3+1-dimensional, so the extra dimensions have to be hidden.

OR. the theory is in a mess?

RIGHT.

Curling up the Extras

The six extra dimensions have to be compactified ...

WHICH MEANS?

... curling them up into a kind of six-dimensional ball. Imagine rolling out a thin layer of pastry; making it into a hollow ball or into a hollow doughnut would be two ways of compactifying in two dimensions.

Because there are specific rules to follow, compactification in six dimensions is **much** more complicated. And there are millions of ways of doing it.

Compactification takes place at each point of space-time. The different shapes and sizes of the compactified space, which is Planck-length-sized and therefore unobservably small, are labelled by parameters called **moduli**. There will be hundreds of moduli, and there are other parameters that have to be chosen to fix a particular string theory space-time. Thus string theory allows an enormous number of possible environments, each of which has its own set of elementary particles and its own laws of physics. These environments form the **Cosmic Landscape** of Leonard Susskind. It is not a real place, but a mathematical construction. The Landscape may have thousands of dimensions. The actual elementary particles and laws of physics in the Universe that we observe are represented by a single point on this Landscape of possibilities. Susskind estimates that there could be 10^{500} or more other possible environments.

What about that eleventh dimension?

Not only were there six extra dimensions and an enormous number of possible environments – it was soon discovered that there were not one, not two, but **five** different string theories. To string theorists hoping that they were on the trail of one master Theory of Everything, this was an embarrassment of riches. Happily, in 1995 the string physicist Ed Witten showed that the five theories could be considered to be different limiting cases of a single eleven-dimensional theory, which he called **M-theory**. M-theory also has a sixth offspring, called **11-dimensional supergravity**, besides the five string theories. The properties of the various string theories are retrieved by compactification.

Little is as yet known about M-theory – not even why Ed Witten called it that. Among the suggestions are Magical, Mystery, Membrane, Master, Matrix, Mother of all theories ...

The basic objects in M-theory are not one-dimensional strings, but two-dimensional objects called (mem)branes or 2-branes, which resemble elastic ribbons instead of rubber bands. Two-dimensional membranes with one dimension compactified play the role of strings.

Other objects with numbers of space dimensions up to nine – 3-branes, 4-branes, and so on – occur. Higher-dimensional branes also occur in the string theories.

M-theory is seriously incomplete. One expert, Brian Greene, remarks that "beyond a hodgepodge of properties, **no one knows what this eleven-dimensional theory is.**" The theory can be applied in many situations, but its exact equations are unknown.

This situation has been compared to the situation of the theories of electricity and magnetism before James Clerk Maxwell unified them in the 1860s. It was apparent that they were related, but not clear what the relationship was until Maxwell published his equations. The M-theorists look forward to a similar revelation.

The eternity of inflation

There is no reason to think that everything that there is in the Universe can be seen from here and now. Although the time since the big bang is only about 13.7 billion years, anything less than about 40 billion light-years away could be visible: the most distant things which we could see have receded because of the cosmic expansion. The observable part of the Universe – call it the **Hubble volume** – lies in a sphere of radius of about 40 billion light-years. Anything outside the Hubble volume is unobservable. But the Universe could in its entirety be spatially infinite. Not only that; processes at locations outside each other's Hubble volume must take place independently of one another. They cannot be causally connected.

When inflation is taken into account, this realization gives rise to some highly speculative possibilities. The process of eternal inflation amounts to eternal self-reproduction of the Universe. As the Universe expands by inflation, quantum fluctuations lead to the production of separate mini-universes, which are causally disconnected from one another and which in turn proceed to expand independently. As Andrei Linde has explained: "The existence of this process implies that the universe will never disappear as a whole. Some of its parts may collapse, the life in our part of the universe may perish, but there will always be some other parts of the universe where life will appear again and again, in all of its possible forms … there will be an end for inflation at any particular point. But there will be no end for the evolution of the universe **as a whole** …" He goes on to explain that eternal inflation divides the Universe into infinitely many large domains that can have different laws of physics. These domains are enormously larger than the Hubble volume – a typical one might be $10^{10,000,000,000}$ cm in size, compared to the radius of the Hubble volume, which is about 4×10^{28} cm.

The collection of independent mini-universes, or pocket universes, is generally called the **Multiverse**. Susskind prefers to call it the **Megaverse**. Whatever you call it, it raises further problems for cosmology, because of fine-tuning.

The fine-tuning problem

From bacteria to galactic clusters, there are complex objects of a huge range of sizes in the Universe. It seems that the Universe has to satisfy a great many highly restrictive conditions in order that these complex structures can exist. Some examples:

The strong nuclear force holds protons and neutrons together in atomic nuclei. It is not quite strong enough to overcome the electrical repulsion between two protons and hold them together in a form called the diproton; if it were only 3.4% stronger, then the diproton would be a stable object and all the hydrogen in the Universe would have been converted into the isotope He^2 of helium in the early stages of the big bang. There would be no hydrogen or long-lived stars in existence today. Nor people. If, on the other hand, the strong nuclear force were 9% weaker, then deuterium, which is the heavy hydrogen isotope H^2, would not be stable, and a key link in the formation of heavier elements needed for terrestrial biology would be missing.

Small differences in the masses of the fundamental particles are also crucial. Neutrons are marginally heavier than protons, and they decay into protons. The neutron and the proton are both more than 1800 times heavier than the electron, but the difference between their masses is only about 2½ electron masses. If instead the proton were heavier, and other relationships were unchanged, then protons would decay into neutrons; in that Universe stars and planets could not exist.

The relationship between the gravitational force and the other forces is also crucial. The gravitational attraction between two protons is about 10^{36} times weaker than the electrical repulsion between them. Suppose, other conditions remaining unchanged, that gravitational attraction were a million times stronger than it actually is. For protons it would still be 10^{30} times weaker than the electrical repulsion. What would be the consequences? Galaxies would form much more quickly, with stars smaller; these stars would typically live for about ten thousand years instead of ten billion years. It would be difficult to form planetary systems, and in any case there would be insufficient time for the evolution of complex organisms. There would be none of us.

BIG numbers — BIG coincidences

There are some big ratios in cosmology. For example, the ratio

$$\frac{\textit{time since the big bang}}{\textit{time for light to cross a hydrogen atom}}$$

is close to 12.3×10^{35}. And

$$\frac{\textit{electric repulsion between two protons}}{\textit{gravitational attraction between them}}$$

is about 12.4×10^{35}.

THESE TWO BIG
NUMBERS ARE
ABOUT THE
SAME ...

... WHICH is partly
a coincidence!
But only partly!—

If you use electrons instead of
protons, the second ratio would be about
4.18×10^{42}, over three million times bigger.
How close the two numbers come
depends on what exactly you choose to
compare. But conventional theory offers
no reason why they should be at all
similar.

Bob Dicke (1961): An accidental correspondence between these numbers would indeed seem very unlikely if there were no reason for any of them to have one value rather than another. But there is a reason: The age of the Universe now can't be too small or too big.

It has to be old enough for elements other than hydrogen to exist — it is well known that carbon is required to make physicists. And it has to be young enough for there still to be a hospitable home for us in the form of a planet going round a luminous star. So the time since the big bang cannot be too small or too big. It has to be comparable to the lifetime of a Sun-like star.

Life in the Multiverse

As far as is known from experiments on and near the Earth and from astronomical observation, the set of elementary particles and the laws of physics are the same throughout the observable part of the Universe – our Hubble volume. The same particles and laws may extend beyond the Hubble volume, and if the eternal inflation scenario, or something like it, is right, continue to hold within a mini-universe formed in a single inflationary event. Then there may be other parts of the Universe – other mini-universes in the Multiverse – where the particles or the laws are different, and as the examples of fine-tuning show, human life, or indeed any kind of complex structure, may be impossible there. However, inflation is such that these other mini-universes are much too far away for us to see them or them to see us.

A speculative means for the formation of mini-universes has been suggested by Lee Smolin. He proposes that new mini-universes could form inside black holes, invisible from the outside, and might then expand and develop, perhaps with laws of physics not very different from those of the parent mini-universe in which they formed. If this were the case, there would be a very large number of mini-universes around and about.

An even more speculative mode of mini-universe formation has been proposed by Max Tegmark. Tegmark remarks that: "If space is infinite and the distribution of matter is sufficiently uniform on large scales, then even the most unlikely events must take place somewhere." If space were not infinite, he wonders, what would lie beyond the sign saying
SPACE ENDS HERE – MIND THE GAP.
From his assumptions, he would conclude that there is another person indistinguishable from you, on a planet indistinguishable from Earth, reading this book, somewhere about $10^{100,000,000,000,000,000,000,000,000}$ meters from here.

In these examples, one expects that there would be distinct causally independent mini-universes with different particle sets and different laws of physics. No doubt other causally disconnected Multiverses can be devised where the mini-universes have the same particles and the same laws of physics.

But is it science?

If there is only one Universe with a unique set of elementary particles and unique laws of physics, then fine-tuning presents a problem: How is it that the Universe is adjusted in such a way that complex structures, including humans, can exist?

At times during the 20th century, it was hoped that a unique physical theory would be found which made everything come out just right, including the possibility of human existence. Inflation and the Cosmic Landscape have left little room at present for optimism about this dream of a Final Theory. There remain at least three options:

      1. Any apparent fine-tuning is a fluke.

      2. There is a Multiverse with a wide assortment of mini-universes, in at least one of which there are conditions consistent with human existence, and trivially the mini-universe where humans are found is one of those.

      3. The existence of a Universe, or of a mini-universe, consistent with human existence is the work of a Great Designer.

Most scientists generally prefer one of the first two options, and regard the third as not scientific. They would concur with Laplace's reply to Napoleon when asked why he had not mentioned God in his book: "I have no need of that hypothesis."

Some proponents of the first option maintain that measures of fine-tuning are not in fact meaningful, because there is no way of assigning probabilities to the different values of the relevant parameters. For example, if the strength of a force can vary over an infinite range, then **any** finite range is only infinitesimally probable. At the same time, a vast literature has been developed in the course of attempts to achieve an assignment of probabilities. This includes attempts to fix probabilities in hypothetical mini-universes causally disconnected from ours, with which there is in principle no prospect of communication. Many critics, employing Karl Popper's criterion that good scientific speculations should at least be falsifiable, dismiss these attempts as bad science. And then there are the anthropicists.

Anthropics

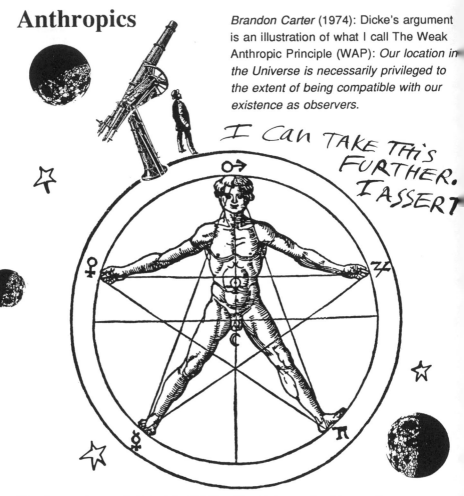

Brandon Carter (1974): Dicke's argument is an illustration of what I call The Weak Anthropic Principle (WAP): *Our location in the Universe is necessarily privileged to the extent of being compatible with our existence as observers.*

I Can TAKE THis FURTHER. I ASSERT

The Strong Anthropic Principle (SAP): *The Universe (and hence the fundamental parameters on which it depends) must be such as to admit the creation of observers within it at some stage.*

What could he have meant by "must"? He seems to have meant that from the existence of humans one could deduce restrictions on the constants of physics. But other people have taken it to mean that the Universe must have been designed so that people could evolve in it. Which leads them back to a belief in a Great Designer.

Some people even believe that if there weren't any conscious humans, there wouldn't be any Universe.

Bishop Berkeley (1710):

John Wheeler (1975): I suggest the Participatory Anthropic Principle (PAP): *Observers are necessary to bring the Universe into being.*

And some talk about the long-term consequences:

John Barrow and Frank Tipler: Here is the Final Anthropic Principle (FAP): *Intelligent information-processing must come into existence in the Universe, and, once it comes into existence, it will never die out.*

Some people still Think...

that the business of science is to find a Theory of Everything, answer "Why?" questions, and mix theology with cosmology.

The eventual goal of science is to provide a single theory that describes the whole universe.... if we do discover a complete theory, then we shall all, philosophers, scientists, and just ordinary people, be able to take part in the discussion of the question of why it is that we and the universe exist. If we find the answer to that, it would be the ultimate triumph of human reason — for then we would know the mind of God.

Stephen Hawking

THIS 'ORDINARY PERSON' ASKS: WHAT IF GOD HAS A CHANGE OF MIND!

Some people don't

People say to me, "Are you looking for the ultimate laws of physics?" No, I'm not.... If it turns out that there is a simple ultimate law which explains everything, so be it — that would be very nice to discover. If it turns out that it's like an onion with millions of layers... then that's the way it is.... Whether or not nature has an ultimate, simple, unified, beautiful form is an open question....

Richard Feynman.

And so...

Cosmology is in an exciting state. Technical developments in astronomy have led to an era of precision cosmology. But to reconcile theory with observation, cosmologists must assume that galaxies are largely made of stuff they can't identify. The Universe expands, but there is no adequate theory of what started the expansion. The best available cosmological models are based on general relativity, but quantum mechanics is needed too, and for over half a century these two theories have resisted all attempts to bring them together. Right now, cosmological theory is in a mess.

In the Middle Ages and after, the theory of planetary motions developed by Ptolemy was in continual need of repair, until Copernicus and Kepler overturned it altogether and started afresh. Today, cosmological theory is in continual need of repair. Attempts to fix it, such as inflation and anthropics, have been unsatisfactory. Perhaps it too needs to be overturned.

Certainly there is no sign of an end to cosmology, or of a "complete" theory. Like other branches of science, it evolves. As long as there are humans, human science will continue to evolve. But human science may not be the only possible science. Only anthropocentric parochialism could allow one to suppose that the Universe needs humans, or, for that matter, to presume that there are no non-human scientists elsewhere, with cosmological theories of their own.

9   I want! I want!

Pub by W Blake 17 May 1793

We are the way we are because
the Universe is the way it is ●●●

●●●and not vice versa

Notes and References

In the references, *The New Physics* = *The New Physics*, edited by Paul Davies, Cambridge University Press, 1989.

References in the form astro-ph/0302131 or hep-th/0503195 may be found at the website arxiv.org

page

8 Richard Feynman's remarks are from his *Lectures on Physics 1*, pp 1-2, 3-6.

13 A verse of "The Love Song of J. Alfred Prufrock" is quoted from T.S. Eliot's *Collected Poems 1909-1935*, Harcourt Brace, New York, 1936.

14 quoted from J.M. Plumley, "The Cosmology of Ancient Egypt", in *Ancient Cosmologies*, edited by Carmen Blacker and Michael Loewe, London, George Allen and Unwin, 1975.

23 Vitruvius, *The Ten Books on Architecture*, trans. M.H. Morgan, Harvard University Press, 1926, pp 261-262, quoted by T.S. Kuhn, *The Copernican Revolution*, Vintage Books, 1959, pp 51-52.

25 C.S. Lewis, *Miracles* (1947), Fontana Books edition, p 82. Harry Emerson Fosdick, *The Modern Use of the Bible* (1924), pp 46-47.

26 Ko Hung quoted from J. Needham, "The Cosmology of Early China", in *Ancient Cosmologies*, edited by Carmen Blacker and Michael Loewe, London, George Allen and Unwin, 1975.

27 Adapted from a translation of the *Mahapurana* (The Great Legend), in *Sources of Indian Tradition*, edited by W. Theodore du Bary, New York, Columbia University Press, 1966, 1, pp 76-78.

31 Luther and Calvin quoted from A.D. White, *A History of the Warfare of Science with Theology in Christendom*, New York, Appleton, 1897, 1, 126.

33 For Kepler's astrology see T.S. Kuhn, *The Copernican Revolution*, p 93.

36 For the prohibition of teaching about sunspots, see A.D. White, *A History of the Warfare of Science with Theology in Christendom*, New York, Appleton, 1897.

36 For Father Clavius's theory of the moon, see Geymonat, *Galileo Galilei*.

40 Galileo's abjuration is translated from *Le opere di Galileo Galilei*, ed. A. Favaro, 19, 406-407, 1909.

42 It is not true, as asserted by Paul Davies, *The Cosmic Blueprint*, Simon and Schuster, 1989, p 10, that Uranus was discovered by its gravitational effects on other planets.

50 Discovery of the bubble-like distributions of galaxies are reported by Valérie de Lapparent, Margaret Geller and John Huchra in *The Astrophysical Journal*, 302, L1-L5, 1986. Discovery of the Great Wall is reported by Geller and Huchra in *Science*, 246, 897, 1989.

54 Fred Hoyle's invention of the term "big bang", and his comment on it, are in his book *The Nature of the Universe*, Blackwell, Oxford, 1950, p 102.

55 The name, Big Bang, is unfortunate ...: Peebles, Schramm, Turner and Kron, *Nature*, 352, 769, 1991. For the limitations of current theory, see "Quantum Gravity", the chapter by Chris Isham in *The New Physics*.

64 The quotations about laws of nature are from Steven F. Mason, *A History of the Sciences*, new revised edition, Collier Books, New York, 1962, p 173.

The comment by Eugene Wigner appears in his article "The Unreasonable Effectiveness of Mathematics in the Natural Sciences", *Communications on Pure and Applied Mathematics*, 13, 1-14, 1960.

68   According to current theory, the proton and neutron in turn are made up of particles still more fundamental, called quarks.

79   The phenomena of emission and absorption are more complex than Kirchhoff's rules suggest. Many different atomic processes may be active in a source of radiation.

82   For more on special relativity, see *Introducing Einstein*.

84   Strictly speaking, only a point has a world-line; an extended object is represented by a world-tube in space-time, consisting of the world-lines of all its points.

103 At velocities small compared to the velocity of light, the formula (for a uniform spherical object) is (escape velocity)^2 = 2GM/r, where M is the mass of the object, r its radius, and G the gravitational constant.

116 Chris Isham's comments in "Quantum Gravity", his chapter in *The New Physics*.

119 Mark Kac's comment on models: *Science*, **166**, 699, 1969.

133 Freeman Dyson's speculations about the far future are published in *Reviews of Modern Physics*, **51**, 447-460, 1979.

134 Extracts from "Spain 1937" quoted from Auden's *Collected Shorter Poems, 1930-1944*, Faber and Faber, London, 1950. Auden omitted the poem from his *Collected Shorter Poems 1927-1957*, published in 1966. Richard Goodman's comment on it in the *Daily Worker* quoted from W.H. Auden: *The Critical Heritage*, edited by John Haffenden, Routledge 1983. For Auden's 1964 view, see *Poetry of the Thirties*, collected by Robin Skelton, Penguin, 1964.

135 The comment on the history of astronomy was made by Malcolm Longair in his chapter "The New Astrophysics" in *The New Physics*.

137 Engels's comment appears in a letter to H. Starkenburg dated 25 January 1894; see Dona Torr, ed., Marx and Engels, *Selected Correspondence*, London, Lawrence and Wishart, 1936, p 517.

146 For more about inflation, see *The Inflationary Universe* by Alan Guth, Vintage, 1998

148 For dark matter, see *The Fifth Essence*, by Lawrence Krauss, Vintage Books, 1990.

152-3 Leonard Susskind explains the Cosmic Landscape in his book of that title.

154 Andrei Linde is quoted from hep-th/0503195

155 Bob Dicke's original paper is in *Nature*, **192**, 440, 1961.

158 Max Tegmark is quoted from astro-ph/0302131

162 For Carter's definitions, see his article "Large Number Coincidences and the Anthropic Principle in Cosmology", in M.S. Longair, ed., *Confrontation of Cosmological Theories with Observational Data*, International Astronomical Union Symposium No. 63, Reidel, Dordrecht and Boston 1974.

163 Berkeley quoted from *Principles of Human Knowledge*, section 44. Wheeler, paraphrased by Barrow and Tipler, *Anthropic Cosmological Principle*, Oxford, Clarendon Press, 1986, p 22. Barrow and Tipler's FAP, op. cit., p 23. Martin Gardner, *New York Review of Books*, 1986, No. 8, p 22.

164 Stephen Hawking, *A Brief History of Time*, Bantam Press, London, 1988, pp 10 & 175.

165 Interviews with Richard Feynman quoted from *Genius: Richard Feynman and Modern Physics* by James Gleick, Little Brown, London, 1992, pp 432-433.

Glossary

absolute zero: the lowest possible temperature, about −273 degrees Celsius; it can be approached, but not attained: 139

acceleration: rate of change of speed: 92, 94

aether: an invisible, weightless, crystalline solid which according to Aristotle fills the Universe, beyond the Earth: 20

anthropic principle: the idea that the place of humans in the Universe is not arbitrary, but must be consistent with the requirement that they should have had time to develop, and that the Universe must be such as to have allowed this: 160

billion: a thousand million; 10

black body: an object which can emit and absorb, but does not reflect, electromagnetic radiation; any distribution of matter in equilibrium with electromagnetic radiation, so that conditions are steady and as much is absorbed as is emitted, behaves as a black body: 110

black hole: an object so small for its mass that its gravitational field prevents the escape of matter or radiation: 103-104; believed to be the final state of evolution of a star of sufficiently great mass; quasars may be black holes: 110

COBE: COsmic Background Explorer satellite: 139-141

comet: a celestial object with a fuzzy head and a long tail which moves comparatively rapidly across the sky. Near the centre of the head there is a "nucleus" of starlike appearance. The nucleus is a small irregular solid object made of silicate dust, ice, and frozen organic compounds. Sunlight spreads some of the dust out to make the tail: 3,32

deferent: in Ptolemy's description of planetary motion: the large circle representing approximately the motion of a planet; the planet itself moves on an epicycle which is carried around the deferent: 24

density: the amount of mass in a unit of volume; the future evolution of the Universe is believed to depend on its average density: 115, 129, 144, 150

deuterium: "heavy hydrogen", an isotope of hydrogen with one proton and one neutron in each nucleus: 69, 127, 150

electron: a charged elementary particle; an atom consists of a nucleus surrounded by a cloud of electrons: 67-70

elementary particle: a basic constituent of matter: 53, 113, 146

epicycle: in Ptolemy's description of planetary motion: a small circle whose centre moves around the circumference of a larger circle; the larger circle may be another epicycle, or the deferent: 24

galaxy: a system of stars, gas and dust held together by gravitational forces: 45-50

homogeneous: having characteristics which do not vary from place to place: 122, 123, 125, 138

infra-red: electromagnetic radiation of frequencies just below those of visible light: 72

isotope: a particular kind of nucleus of an element with a particular number of neutrons: 69

isotropic: having the same appearance in all directions: 122, 123, 125, 138, 147

Jupiter: the largest planet in the Solar System: 23, 34, 35

light-year (ly): the distance light travels in a year, which is about 9.46 thousand billion (9.46×10^{12}) kilometers: 7

mass: a measure of inertia, which means resistance to being accelerated; 92

Megahertz (MHz): a unit of frequency = one million Hertz, or one million cycles of oscillation per second: 75

Mercury: the innermost planet of the Solar System: 95

nebula (plural: nebulae, from the Latin word for "mist"): originally used to describe fuzzy or cloud-like celestial objects; those in which, with improved telescopes, stars have been distinguished, are now generally called "galaxies", while "gaseous nebulae" refers to clouds of interstellar gas and dust within this Galaxy: 42-45

neutron: an uncharged particle, one of the constituent particles of atomic nuclei, the other being the proton: 68

neutron star: a celestial object composed entirely of neutrons; believed to be the final state of evolution for a star of moderate mass: 108

nova (plural: novae, from the Latin word for "new"): a star which over a short period increases greatly in brightness: 32

nucleon: a neutron or proton: 69

nucleus (plural: nuclei, from the Latin word for "nut"): the central part of an atom, comprising nearly all of the mass: 67-70

Planck density: a density, about 10^{93} times the density of water, constructed from the fundamental constants of nature: 55, 115

proton: a charged particle, one of the constituent particles of atomic nuclei, the other being the neutron: 68

quantum (plural: quanta, from the Latin word for "how much"): a packet of energy: 110

quantum mechanics: theory of motion of material objects, especially relevant to systems of atomic size and smaller: 66, 113

quasar: quasi-stellar radio source: 110

radiation, electromagnetic: an oscillating electromagnetic field, propagated in space; forms of electromagnetic radiation include visible light, radio waves, infra-red and ultraviolet radiation, X-rays and gamma rays: 72-81

radiation, gravitational: an oscillating gravitational field, propagated in space: 100

red giant: a star larger and cooler than the Sun; the Sun is expected to evolve into a red giant: 106

relativity theory: (1) special relativity: theory of motion of material objects, especially relevant when there are relative motions at speeds comparable to the speed of light: 82-87; (2) general relativity: theory of motion of material objects, especially relevant on the scale of planetary systems and upwards and whenever gravitational fields are of prime importance: 66, 95-98

solstice: one of the two times of year when the Sun is highest or lowest in the sky: 16

space-time: in relativity theory, space (three-dimensional) and time (one-dimensional) considered together as the four-dimensional arena where physical processes take place: 82

supernova (plural: supernovae): a star which over a short period increases in brightness by a factor of a billion or more: 32, 107-110

ultra-violet: electromagnetic radiation of frequencies just above those of visible light: 72, 137

white dwarf: a small, dense hot star; the Sun is expected to evolve into a white dwarf after its red giant phase: 106

Biographical Index

Gardner, Martin (1914–) US science writer: 161

Geller, Margaret (1947–) US astronomer, co-discoverer of the "Great Wall": 50

George III (1738–1820) King of England from 1760: 43

Gold, Thomas (1920–2004) Austrian-born US astronomer: 56

Goodman, Richard (1911–66) English poet: 134

Greene, Brian (1963–): US physicist: 153

Guth, Alan (1947–) US physicist; inventor of inflationary Universe: 145-146

Hawking, Stephen (1942–) English physicist; invented theory of radiation from black holes: 103, 162

Heisenberg, Werner (1901–76) German physicist; inventor of one form of quantum theory: 66, 113

Herschel, Caroline Lucretia (1750–1848) English astronomer, sister of William: 42-43

Herschel, William (1738–1822) English astronomer, discoverer of the planet Uranus: 42-43

Hertz, Heinrich (1857–94) German physicist: produced radio waves and showed that they move with the same speed as light waves: 74-75

Hipparchus (fl. 160 BC–125 BC) Greek astronomer: 24

Hoyle, Fred (1915–2001) English astrophysicist: 54, 56

Hubble, Edwin (1889–1953) US astronomer; discovered the expansion of the Universe: 45, 51-52, 60, 118-119, 154, 157

Huchra, John (1948–) US astronomer; co-discoverer of the "Great Wall": 50

Isham, Christopher (1944–) English physicist: 116

Jansky, Karl (1905–50) US physicist; discovered radio-frequency radiation from Galaxy: 136

Jinasena, 9th-century Indian religious teacher: 27

Kant, Immanuel (1724–1804) German philosopher: 42

Kaluza, Theodor (1885–1954) German mathematical physicist

Kepler, Johannes (1571–1630) German astronomer; formulated the laws of motion of the planets named after him: 33-34, 164

Kirchhoff, Gustav Robert (1824–87) German physicist: 79

Klein, Oskar (1894–1977) Swedish mathematical physicist: 150

Ko Hung (3rd–4th century AD) Chinese alchemist and astronomer: 26

Laplace, Pierre Simon, Marquis de (1749-1827) French mathematician: 159

Lewis, Clive Staples (1898–1963) Irish-born writer: 25

Linde, Andrei (1948–) Russian-born physicist, working in the USA: 154

Luther, Martin (1483–1546) German theologian: 31

Maxwell, James Clerk (1831–79) Scottish physicist who formulated the equations of the electromagnetic field, predicting the existence of electromagnetic waves and identifying light as one form of these: 73, 82, 102, 153

Napoleon Bonaparte (1769–1821) emperor of France: 159

Newton, Isaac (1642–1727) English physicist and mathematician; invented the theories of mechanics and of gravity named after him: 41, 82, 90-95

Osiander, Andreas (1498–1552) German theologian: 29

Pappus of Alexandria (4th century AD) Greek mathematician: 172

Peebles, (Phillip) James (Edwin) (1935–) US astrophysicist: 55, 126

Other Books to Read

Besides the books referred to in the Notes, here are the names of a few others, out of the hundreds in the bookshops, which you may find interesting or useful.

Greene, Brian, *The Elegant Universe*, Random House paperback, 2000. Non-technical introduction to string theory. ISBN 0375708111

Guth, Alan, *The Inflationary Universe*, Vintage, 1999. ISBN 009995950X

Lightman, Alan and Brawer, Roberta, *Origins. The Lives and Worlds of Modern Cosmologists*, Harvard University Press, 1990. Interviews with 27 cosmologists, including Geller, Guth, Hoyle and Hawking.

Rees, Martin, *Just Six Numbers*, Phoenix, 2000. ISBN 0753810220

Smolin, Lee, *Three Roads to Quantum Gravity*, Phoenix, 2001. ISBN 0753812614

Felix Pirani is Emeritus Professor of Rational Mechanics in the University of London. He has published a number of scientific papers and popular articles about Einstein's theory of relativity, and three books for children. He was once attached to the group of cosmologists known as the Cambridge Circus.

Christine Roche is a French-Canadian cartoonist/illustrator who lives and works in London. She draws for various magazines and has illustrated several adult and children's books. Her books include *I'm not a Feminist, but...*, a book of cartoons, and *A Woman's History of Sex* (with Harriet Gilbert).

Felix Pirani and Christine Roche (neither of whom owns a dog) have collaborated on three other books. One of these was condemned by a Parliamentary motion for inciting children to alcoholism and violence.

Acknowledgements

Thanks to Richard Appignanesi, Monika Goppold, Bianca and Marta Monteleoni, J Ravetz, David Robinson and Chris Roper for their comments and to Frank Weil for his unstinting encouragement.

Index